Against All Odds

By

Cecelia Rose

Published by Barnes and Noble Publishers.

Cover design by Barnes and Noble Publishers.

ISBN: Printed in the United States

Barnes & noble Publishers

Note from Author

If you find yourselves in any parts of this fictional story, it is with great hope that this story may offer some form of emotional healing. Please understand that it's imperative that you learn to love yourselves regardless of how others may have acted towards you or may have treated you! It is with hope that you understand that no one can or will ever love you more than you should ever love yourselves.

Preface

This story is about the dysfunctions of a family who creates a scapegoat, however, what's even more sad is that it's with one of their very own. The chosen scapegoat will learn that she'll be ostracized by the people who are supposed to love… support… and protect her when no one else will. The chosen scapegoat will grow up learning that she'll be made to feel that she has never had a place as part of the family with her relatives. This chosen scapegoat will also be made to feel more like an unwanted guest in-regards to her relatives. The chosen scapegoat will struggle to endure a life of lies…hurt… and distrust. However, she'll learn that she's been guided by a spirit of overcoming! Fortunately, the chosen scapegoat will also embrace her inner strength aided by the overcoming spirit. The overcoming spirit has been set in place for her to be able to withstand and endure all the schemes conjured up for every demise planned against her.

The chosen scapegoat will watch and listen as both the parents prop up the pawns and will be praised over the smallest of things for any and everyone who will listen. The family will leave the chosen scapegoat feeling insignificant by them all. However, the chosen scapegoat will learn a lot about herself while going through life looking… seeking… and searching for love. There will be one thing the scapegoat will eventually figure out and that is with each attempt, there will come plenty of fails before she realizes her worth!

The chosen scapegoat then understands that it will all be due to her trying to fill an emptiness that she had been made to feel from each previous empty void that crossed her path, but she'll deeply realize that those voids taught her the most valuable lessons of life! The chosen scapegoat will rise a new realizing that the junk she allowed she could not make into any promise.

Now, what the family will never understand regarding the chosen scapegoat is how she keeps enduring because what the family doesn't know is that it's due to what has been planted inside the scapegoat before birth.

Table of Contents

Character Introduction

Role Characters:

The Narcissists -The Cultivator - The Pawns - The Scapegoat

The **mother** is the **narcissist** of this family, and she is the one who plants all the seeds concerning her daughter, the **chosen scapegoat,** and for all the chain of events that will soon come her way.

The **narcissistic mother** then immediately appoints all the players in their starting lineup positions! The first player is the **cultivator**, the chosen scapegoat's **father**. The cultivator is the one with whom the narcissist plants most of her seeds, mainly if she wants to hide her hands after she's already thrown the rocks.

The next players are the **chosen scapegoats**, two **siblings**, also known as the **pawns**, these pawns will be made to feel privileged constantly, and each player will make sure to carry out whatever snare that is assigned to them.

Usually, all assignments are orchestrated by the **narcissistic mother** and at the finish of each scheme, these players will also have what is known as a glory effect! The glory effect is celebratory after the snares have been completed. That is when they'll all sit back to snicker amongst themselves about the wicked deeds they've enjoyed regardless of how the **chosen scapegoat** had been left made to feel.

The **chosen scapegoat** from birth, and as she grows older, will be treated indifferently by both parents and siblings; however, the one thing she is going to learn is just how blessed she is with the gift that only the "**Most High**" can provide! That planted gift will cause her to be able to see all who are meaning to invoke all sorts of evil upon her. The chosen scapegoat will eventually overcome those rejections, but more so from her family! Eventually, her battle scars, as she had known, will no longer be disturbed by anymore foolishness.

The **narcissist** manipulates everyone in the family where it's always continuous chaos through the controlled positions, and each player knows their roles, which they will play very well; however, if any of the **players** were ever to get caught in their schemes, the end results were to play the victim. Each player in their role also understood that they were never to apologize or accept any responsibility for their actions when it came to the chosen scapegoat.

Now, the players have a delusion of grandeur, believing that the **chosen scapegoat** should remain loyal to them regardless of how they bruise or hurt her.

Chapter 01: The Cultivator Sabastian Bouvier And the Bouvier Family

Sabastian Bouvier, the cultivator, is better known as Bash to all who know and love him.

Bash is a 6'11 tall, slender man with an easy, meek personality, and he always seems slow in anger. Bash is very much like his father in that regard, and he's also the man whom Bash realized he wanted to pattern himself after, at least when it comes to family.

Bash's complexion is of a radiant warm brown tone, much like smooth caramel. Bash has these dark eyes that give an illusion of twinkling; his eyelashes are long and dark, which gives him a mesmerizing, dreamy look, but when he smiles, those pearly whites can brighten up any room.

Bash mostly takes his good looks from his mother but his strong, well-put-together body from his father. Most women consider Bash a showstopper, and when the ladies see him, all they can do is gaze at him with lustful desires.

Bash looks like a chiseled piece of art where most women can only hope for an opportunity that he might be interested in them. However, unknowingly, the one thing that the women didn't know was that Bash had already set his sights on one woman, but he wouldn't be single for very much longer.

Bash's father "Jacques Bouvier"

Jacques is this 6'2 tall, strapping, muscular-built man who also looks like a chiseled piece of art. Jacques has a reserved, quiet-like demeanor because he listens more than he talks! Jacques carries himself like a man with great dignity and with confidence.

Jacques is a well-dressed man on a daily, and he's also in his last year of law school. Jacques has already begun to wear tailor-made suits to prepare himself for the role he is about to embark upon. Jacques understood what it was going to take for him to stand on the side of justice, but more so for a man of color! Jacques knew that he had to set himself apart. His first goal was to pass the bar exam on his first attempt immediately following his graduation. Jacques's career was not the only thing he was setting goals on and preparing to have; he was also preparing for his fortune with the most beautiful girl on the campus; you see, he wanted the best of what this world had to offer him.

Jacques knew that when it came to marriage and plans of having a family, he set his sights and heart on the campus's beauty.

Bash's Mother "Madeleine Stricklin-Bouvier"

Madeleine is a 5'3 petite, curvy woman with bright caramel-colored skin. She has large dark eyes and a smile that resembles that of the Mona Lisa. Madeleine has thick, dark, beautiful long locs that grew down to her behind; however, what stood out the most about Madeleine was her sweet, loving spirit.

Madeleine's loving spirit was the one thing that caught and was going to steal Jacques' heart! Madeleine is also working on her master's degree in education, and her driving passion is to teach children well and to one day teach her future children. Madeleine had begun to spend more time with Jacques, and she, too, wanted to be a part of his life with hopes of marriage and to have a family with him!

When it comes to the other men on campus, when they saw Madeleine, they may have long to know her, but her heart will, by all means, remain true to only one man, and that one man is Jacques Bouvier.

When Jacques wasn't with his head stuck in his law books or working at the nearby campus library, he was finally spending most of his time with Madeleine. After about a year of dating, they both grew to adore each other. Jacques and Madeleine realized that they were in love and they were headed in the same direction. The couple realized they were so perfectly matched because, at times, they sometimes seemed to mimic or mirror each other when they were together. Madeleine understood that she had stolen his heart and, in return, he had stolen hers.

Jacques, without any doubt, knew that he had found his wife and the mother of his future children. He also knew that it would not be long before he would ask Madeline for her hand in marriage! Jacques had been saving up for a ring that he was going to purchase as the engagement ring for his beauty.

The ring Jacques was about to purchase was one beautiful 2-carat heart-shaped diamond set in a thin band of gold. One evening prior to their graduation, Jacques had decided that this would be the day he would pop the question to his beauty. Jacques took Madeleine out for dinner as they strolled alongside the water's edge, where they often frequent dinners together and the place where they both enjoyed their long romantic walks together! The water's edge was the one place they loved watching the moon caress its beauty on the water, along with watching the palms swaying in the wind.

The water's edge was also a place where time seemed to hold no measure once they were there together, and when the couple came to their favorite spot alongside the water's edge, that's when Jacques said just a second, baby, I need to lace up my shoe; he unhooked his arm from Madeleine's as he bent down resting on one knee waiting for his beauty to take notice of him eventually.

Madeleine was still watching the palms swaying until realizing that it was taking too long for Jacques to be lacing up his shoes. She looked over her shoulder to see why he was taking so long. That was when she noticed Jacques on one knee with a ring box in hand. He looked up at her with such nervous energy and a big smile; he then took Madeleine by her hand as she looked shocked with widened eyes. Jacques said, "Madeleine Stricklin, you are the love of my life; I live and long for you daily, and I can't imagine my life without you. Madeleine, would you please marry me and be my wife?"

Madeleine was utterly speechless for a minute; she hurried and pulled Jacques up from his knee to embrace him; she hugged and kissed him all over his face! Madeleine said, "Yes, of course, I will marry you, Jacques; I've prayed for this day, and I would love to be your wife!"

Jacques wrapped his arms around Madeleine's waist, pulling her closer to him, and then said, "You have made me the luckiest man alive!" Madeleine said, "Monsieur Bouvier, know that I love you through eternality!" Jacques chuckled and then said, "There is no other like you, Madeleine. Believe me, I know that the "Most High" has just blessed me!"

Jacques and Madeleine both resided in New Orleans, where they had grown up in the 9th ward. However, New Orleans at this time was a financially struggling environment after a major hurricane had ripped the city to shreds! The couple knew regarding both their desires that they would hinder them of their dreams, so they agreed that they needed to leave New Orleans and relocate to a larger state with a big city.

Jacques and Madeleine didn't believe in wasting time, so shortly after graduating, they decided to get married; they had a small, intimate outdoor wedding ceremony at the library's flower garden, where Jacques worked with only their immediate families present.

Neither of these families had the money to support or spend on any grand wedding celebration for them, and Jacques had spent his savings on buying Madeleine's ring; however, besides that, a grand ceremony didn't matter to either of them. They knew that the marriage mattered the most to them both; however, Jacques made a promise to Madeleine during their reception. Jacques told her, "One day, once I'm making a partner's salary, I'm going to give us the wedding of our dreams!

Jacques and Madeleine immediately put their minds together regarding their future dreams of prosperity; then, they decided to relocate to Chicago, IL. The newlyweds both knew that there would be more opportunities for them both in a larger city, and that's why they decided upon Chicago despite not having any family there.

Jacques passed his bar exam on his first attempt. He then submitted his resume to several law firms around the Chicago area. He had spoken with one of the leading law firms in Chicago, where he had hoped to work in his field of practice.

Soon after Jacques submitted his resume along with a phone interview, he received an offer for an entry-level law practicing position from the largest Law firm that he had sought out first. Jacques was excited to let his wife know that their dreams were coming to light!

Madeleine was happy for her husband. She immediately encouraged his decision to accept their offer. Jacques knew it was a once-in-a-lifetime opportunity because he had already done his research on that firm before his graduation, and it was the main law firm that he had set his sights on.

As soon as they relocated to Chicago, Madeline landed her first teaching position, and the position was at one of Chicago's elementary schools. Madeleine became one of their new 2nd grade teachers, and how excited she was to be teaching young minds. After a couple of months of teaching, Madeleine fell ill one morning; Madeleine thought she had possibly caught a bug from one of the children; each day thereafter, she noticed being sick only in the mornings; she quickly realized that she and Jacques might be expecting a baby.

Madeline decided to visit a nearby clinic early one afternoon to confirm her suspicions. Once she had seen a doctor, that's when she began discussing her symptoms with the doctor, and he decided that he needed to give her a pregnancy test after a physical exam. A few minutes had passed from her taking the pregnancy test when it came back confirming just as Madeleine had suspected, and the doctor walked back into the exam room. He said, "Congratulations, Mrs. Bouvier; you are pregnant, and you are approximately eight weeks." Madeline felt overjoyed. After the visit, she slowly drove home, thinking of how she wanted to reveal to Jacques their great news.

Madeleine stopped at a store on her way home to get some food items, and once she made it home, she began preparing her husband a surprise dinner with everything she knew he loved. Now, with the excitement, she was feeling like she could hardly contain herself; she badly wanted to call to inform Jacques about the pregnancy, but she finally chose just to wait until he arrived home. Madeleine couldn't wait to see his face once she told him the news; she knew he would be just as excited as she was!

Jacques arrived home from a long day at the firm; however, he was always eager to see his wife once he arrived; he came in and set down his briefcase before he called out to her, saying, "Baby, I'm home!" Jacques headed towards the kitchen, where he always found his wife each evening preparing dinner once he arrived.

Jacques walked up to his wife wraps his arms around her waist as he snuggled in close, giving Madeleine soft kisses on her cheeks! This is what Jacques did daily because he always liked to be reminded of why he loved his wife, and it seemed to cause him to fall in love a new with her all over again each day.

Jacques said, "It smells so good in here. What are you preparing?"

Madeleine said, "This is a Cajun surprise dinner just for you!" Jacques said, "Oh really! Well, let me go to take a quick shower and freshen up for this surprise dinner," as he hurried away! While Jacques was in the shower, his wife quickly set the table and waited for his return. Jacques returned to the dining room feeling fresh and completely relaxed for his surprise dinner! Jacques slowly strolled around the dining room, then out of curiosity after seeing the table, he said, "Well I see the table is set extra special. However, I do know that it's not our anniversary!" Madeleine smiled. She then rushed to say, "No, it's not our anniversary, honey!" Jacques asked, "So, why have you set the table like this? Madeleine said, "Because it's a very special occasion just for you!" Jacques looked even more puzzled. He said, "Well, at least let me pour us some wine for this surprise dinner and for all of this delicious-looking food."

Madeleine could not wait another minute before she gave her husband another clue; she said, "Only milk for me, Jacques…!" Jacques, with a surprised tone, cluelessly asked, "What is really going on because when have you ever drunk milk for dinner, which is a first?"

Madeleine got up. She guided her husband to his seat and set him down! Madeleine took a seat in his lap, and she softly kissed Jacques's lips before telling him the good news. While looking into Jacques's eyes,

Madeleine said, “Honey, the reason it’s a first and there will be no more wine for me or at least until after our baby arrives!”

It took him a few minutes for it to register for what his wife had just told him, Jacques’s eyes bulged forward with surprise; he began to grin from ear to ear as the tears were building in his eyes! Jacques tried speaking, but he could not get his thought process together. He said to Madeleine, “Wait, so are we; no, wait, that didn’t come out right,” while shaking his head, “I mean, are you pregnant?”

Madeleine smiled big and then said, “Yes, honey, we are about to be parents!” Jacques became almost speechless… all he was able to get out was, “When?” Madeleine said, “In 7 more months, I’d be giving birth to our first little Bouvier baby!” Jacques softly pressed his face into her bosom, hugging her!

Seven months later, a baby boy was born whom they named after his father, naming him Jacques Bouvier Jr! Jacques and Madeleine were enjoying their new addition to their family and watching him grow ever so quickly. Each year after his birth, it was one more Bouvier baby after another.

Jacques and Madeleine didn’t believe in wasting time as they wanted to have their children close in age; they eventually ended up having six children in all. The family ended up with four boys and two girls; the firstborn was Jacques Jr.; a year after Jacques Jr. came Andre; a year after Andre came Kingston; a year after Kingston came Gia; and a year after Gia came Angela. Madeleine then decided that she wanted to wait on having more children for a while because Madeleine’s schedule had become very full; nevertheless, eight years after the birth of Angela, the Bouvier's were blessed with their last son, Sebastian, the baby of the Bouvier family who they all called Bash for short.

One evening, Jacques arrived home from the office. He then walked into the kitchen to fall in love again with his wife, but this evening, when Jacques approached his wife, he noticed how tired Madeleine appeared. Jacques asked her, saying, "Baby, are you alright?" with a concerned tone.

Madeleine smiled, looking up at Jacques with love in her eyes; she said, "Yes, honey, I am fine; please don't make such a fuss. I'm just a little tired! I think I may have just taken on too much way too fast! I think it's probably from a combination of my teaching and serving on that board that I'm on with the school and tutoring our children each day that's draining me."

Madeleine told her husband, "I think that maybe I'll scale back on some hours, and I'm probably going to step down from the board that I serve on for a while. I think if I were to switch over to teaching kindergarten, I could get more rest; besides that, it's a part-time position, it would also leave more time for me to be home with our children when they get home from school and for us to be able to spend more quality time together!"

Jacques agreed with his wife, and he quickly stated, "I think you should really consider doing that because I don't want you overworking yourself, and I don't like seeing you this exhausted." Now that Jacques's career was at its height with a record for winning a majority of his cases, he had become highly sought after for his legal expertise. Jacques had plans to become a partner soon. It has been three months since he expressed his interest in the open partnership role with the firm. Jacques felt comfortable that the partnership with the firm would go exactly as planned, all based on his courtroom winnings, which he knew would substantially increase his salary.

Jacques was hoping to be in that role within any day, and it would allow him to take on certain cases and to be able to groom the new hire attorney's coming into the firm. Jacques wanted to be able to support Madeleine and his children comfortably.

Each evening when Jacques arrived home, he also kept noticing how his wife seemed overly tired. He also figured it must be from his wife having such a loaded schedule as they had already previously discussed.

Jacques's thoughts consistently stayed on his wife after seeing how tired Madeleine kept appearing to him; he really wanted his wife to completely stop working and stay home with the children because he had become very concerned for her health.

Jacques went into the office one morning to speak with the Firm Partners regarding the open partnership role. Upon Jacques's arrival, he asked his paralegal where the partners were. Michelle directed him to go into the conference room where they all were; he walked into the conference room and then greeted everyone. He took a seat and then stated that he needed to address his interest and concern for the opening of the partnership role. The head of the law firm said, "It's interesting that you came to discuss this with us this morning because we were just having the discussion, and we were just about to call you into our meeting to offer you the role of Partner with us."

One of the head Partners spoke up and said, "Jacques, we all came to an equal vote knowing that you're the best candidate as a partner with us; we've seen your dedication and hard work, which has been outstanding! Jacques, we want you to know you've helped to grow our firm to a level that surpasses all others with each one of your winning cases. Jacques, we hope you will accept our offer to become our next partner with our law firm!"

Jacques quickly said, "I am honored to take on this new role. Now, let's discuss this salary increase," as they all chuckled, knowing that Jacques was not one to waste time or to ever miss a beat for being direct.

Jacques was very excited to return home to surprise Madeleine and let her know that she had the option to stay home if she chose to. Jacques arrived home and then walked into the kitchen where he always found his wife; once he arrived home each evening, he eased in and then softly clutched his wife's hand to lead her away from the kitchen.

Madeleine said, "Honey, I have not finished preparing dinner yet!" Jacques spun around, cutting off the stove. He then led his wife away from the kitchen. Jacques led his wife into their bedroom, and he took a seat on the side of their bed. As he began to tell his wife the news, he said, "Come here, Baby, lie down beside me and let me hold you close to me for a little while. Oh! And by the way, no worries, baby, I'll also order dinner tonight. Let's just call it a surprise dinner."

Madeleine said to her husband, "Wait, not another baby surprise, is it?" as she laughed. Jacques smiled and began lightly kissing his wife's forehead! Madeleine then stated, "I see that you're in a love-making mood early this evening, huh?"

Jacques softly chuckled and smiled. He said, "Baby, you know that I am always in a love-making mood with you regardless of the time of day." Jacques said, "Madeleine, please know that without you, I don't think that I could make it!" Madeleine asked her husband, "What are you talking about, Jacques, because I am right here, honey?"

Jacques said, "I can't take another day seeing you this tired baby; I really want you to remain at home until you are feeling better! Madeleine, I know that teaching is your passion, but until you are better, I want, no wait! I mean the children and I, we need you much more!

Madeleine's husband said, "Listen, I was made partner today, which means that the salary is more than enough to greatly sustain our family." He then asked Madeleine, "What do you say?" Madeleine said, "This is great news, and congratulations, honey. I know you have worked diligently towards that role and besides that, right now, I am just too tired to debate about it."

They both laughed at her response. Madeleine then lay on her husband's chest until she quickly drifted off to sleep. Jacques felt as if something was very off regarding his wife; however, he was hoping that if she could receive adequate rest, she could return to being her vibrant self again.

After several months of extreme exhaustion continued with Madeleine, one night, when Jacques arrived home from the firm, their children greeted him at the front door and began explaining to their father with full excitement that they made sandwiches for dinner. Jacques praised the children, and then, in a rush, he asked them, "Where's your mother?" Gia said, "She was in bed asleep." Jacques immediately headed down the hallway to their bedroom to check on his wife.

Jacques had not ever seen his wife in the kitchen at that time of the evening, once he walked into the bedroom, there was where he found Madeleine tucked away in their bed. Jacques woke her up from her sleep. He asked his wife what was going on with her and then asked how she was feeling. Madeleine told her husband that she had been there the entire day due to feeling completely exhausted and that she had told the children to fix themselves sandwiches until he arrived home!

Madeleine also explained to her husband that she believed that maybe she had caught a virus, probably from one of the children at school, right before she had taken on her leave of absence because none of their children were sick.

Jacques decided right then to call to discuss his wife's symptoms with one of his clients, who happened to be a physician. Jacques reached Doctor Daniels, and he began explaining to the Doctor that his wife had been extremely tired for the last few months, and he had found her that evening where she could not even get out of bed.

Doctor Daniels said, "It could be that she just needs some iron; I see this often in women, mainly after women have given childbirth." Jacques told the Doctor he understood, but their youngest child was about to be three years old. He then asked Doctor Daniels if he could make it a point to see his wife as soon as possible because she didn't look well. Doctor Daniels said, "Absolutely, either you or Madeleine. Please call my office first thing in the morning, and I'll speak with my nurse to have her schedule an appointment for your wife right away." Doctor Daniels said, "I'll also schedule a series of tests just to be on the safe side of anything else, but I'm pretty sure that it's nothing that can't be corrected, so rest your minds!"

The next morning, Jacques told his wife that he had called one of the physicians he manages through the firm! Jacques went on to tell Madeleine that Doctor Daniels said her iron could be low, which can make her extremely tired. "Doctor Daniels also stated that he wanted us to call and schedule an appointment for you. He also mentioned that he was going to schedule some other test just to rule out anything else. Jacques went on to say that he said it's probably not something that can't be treated.

Madeleine said, "Honey, I have been taking iron; however, I will give them a call this morning, but as I said, it's probably just a virus." Madeleine called and spoke with the doctor's nurse, who scheduled a same-day appointment later that afternoon.

Madeleine called to let her husband know that she had an appointment scheduled that afternoon, and she needed him to take her. Jacques canceled some appointments first, and then he left the office immediately to take his wife to the appointment. Doctor Daniels walked in to examine Madeleine, but he became visibly concerned upon the exam that Jacques had taken notice of. Madeleine displayed other symptoms, such as bruising with some swollen lymph nodes. Doctor Daniels immediately ordered some cancer screening labs. The doctor was not alluding to his suspicions because he did not want to alarm the Bouviers of what he suspected too soon.

Four days later, Madeleine's results came back, showing that Madeleine had an aggressive form of leukemia. Doctor Daniels knew that this sort of news was going to rock their world hard. Doctor Daniels, on many occasions, had heard Jacques often speak of his wife and how much he loves and adores her.

Chapter 02: Until Death

Doctor Daniels called Jacques at the firm after receiving the results, telling him that he needed to see him and his wife in his office that afternoon! Jacques said, "It sounds serious! Please tell me, is it serious enough that we need to be there today?" Doctor Daniels didn't hesitate with a serious tone as he said, "Yes, Jacques, it is, but we will discuss everything regarding Madeleine's diagnosis once you both are here in my office."

Jacques's mind began racing after their conversation; he then called his wife to tell her that Doctor Daniels needed to see them in his office that afternoon. Madeleine, unaware of the seriousness, asked her husband to give her an hour to get dressed. Jacques said, "Baby, I'll be there shortly to help you just in case you need it."

The Bouviers arrived to see Doctor Daniels, and once he walked in and opened the report, then began explaining the lab diagnosis, the Bouviers were shaken to their cores; meanwhile, Jacques felt helpless as his wife leaned in on him sobbing.

Jacques tried to convince and comfort his wife by telling her that everything was going to be alright, but it was at that precise moment that Madeleine realized what she thought was no more than fatigue; she had let it go on for way too long!

Each day after finding out what was wrong with Madeleine, Jacques would rise from his sleep earlier than usual; he saw to it that his wife had everything she needed or that she could have wanted! Jacques was also tending to their children, a change their children had taken notice of but didn't understand yet what it was.

Jacques and Madeleine finally decided to sit their children down to talk with them about a week later; they both wanted to explain to their children what their mother was dealing with.

Jacques said, "Listen, children, we want you all, but even more so, we need you all to help with everything around the house." Jacques also said, "Which includes loving on your mother as much as you like!" Jacques went on to explain to the children that their mother was fighting an aggressive form of cancer and that they needed to work together for her sake.

Jacques made sure that they knew that their mother needed a lot of rest; he asked their children to please keep as much confusion and bickering down inside of the house so that their mother would be able to rest peacefully! Jacques asked the children can you all please do this for your mother and for us all?

The Bouvier children all nodded yes in agreement but had looks of disbelieving fear on their faces! The children did not want to ask too many questions after that because they also saw the same look of disbelieving fear on their father's face.

Madeleine made sure to speak with her children daily, telling them to take care of one another and to make sure they know that family is the most important thing that you can ever have in life. The Bouvier children's mother would constantly tell the children how blessed she was to have each one of them! Bash, who was just a toddler, always climbed into his mother's lap or into Gia's lap during their talks; at times, he chose Gia's lap because she had a lot of ways, much like his mother. Gia seemed to understand her mother's wishes more than the rest of her siblings, and she picked up the baton immediately.

Weeks had passed, and without a bone marrow match found, Madeleine was steadily growing weaker. Jacques immediately started scaling back hours from the firm to be there to help his wife and their children.

Jacques bathed his wife daily if she was able to withstand being up, and he made sure to cook for her and their children; most nights, once everyone was fast asleep, he was somewhere off to himself praying and talking to the "Most High"!

Jacques was praying with an expectation that his wife might be spared, but he was growing desperate; he had begun bargaining praying! Jacques was tearfully praying hard one night, stating that he would give up his life to trade places with his wife. Jacques understood that this was the one thing he could not negotiate, unlike the laws he was accustomed to, and realized that he had no power to help with his wife's health issue.

Doctor Daniels called Jacques at the firm, requesting that he come to his office early one morning, stating that he needed to speak to him regarding Madeleine's latest pathology results. Jacques, without hesitation, said, "I'm on my way!" When he arrived at the medical office, and Doctor Daniels walked into the exam room to speak with him again, he saw that Doctor Daniels had the same look on his face as when he had first read Madeleine's original diagnosis.

Doctor Daniels said, "Jacques, I wanted to speak with you before I contacted Madeleine. I need you to be able to internalize what I'm about to explain to you," Doctor Daniels nervously continued, "I have the latest lab results from the pathologist for Madeleine. However, unfortunately, Jacques, the results are showing that Madeleine is at the level of stage 4 with this cancer, and a bone marrow match is not looking promising enough at this time to be able to save her."

Jacques's response was, "Wait, what?" Jacques could not process what he had just been told. He said, "Wait, hold on… no… no… no… this can't be right; there must be something else that can be done!" Jacques's eyes were turning flaming red, and then suddenly, the tears began flowing uncontrollably down his face while he was pleading to hear anything else to save Madeleine.

There was no way that Jacques was accepting the fact that he was not about to have the love of his life and the mother of his children. Right before Jacques left the medical office, he asked Doctor Daniels to please let him be the one to inform Madeleine regarding her latest pathology results. Doctor Daniels told him that he would give him one week, and then he would need to tell her himself! Jacques left the medical office, and his car parked in the parking lot while he decided to take a long walk to think.

Jacques thought he could figure out what he needed to do as far as trying to help his wife; however, it was at that precise moment that Jacques realized that he was a lost man! After a while, he arrived back home. The children greeted their father as soon as he arrived. Jacques saw fear in their young faces, and he also understood that he had to not let his emotions show through; he put on a smile and then asked the children if they were hungry.

Gia quickly spoke up, saying, "No, I made soup and sandwiches," but Jacques noticed that Gia never even looked up while speaking to him while holding Bash in her lap! Jacques then hurried and went to check on his wife as usual. Madeleine was asleep in their bed. Jacques got into bed with his wife fully clothed to keep him from feeling defeated! Jacques needed to breathe in her scent, trying to remember everything he could about his wife as he held her close to him.

The leukemia was nothing of a respecter towards Madeleine, and she was fighting trying to stay alive for her loved ones; early the following morning, Madeleine asked Jacques to please give her a bath; he ever so gently carried his wife to a warm tub of waiting water! Jacques had added some minerals, milk, and essential oils to give her a soothing bath.

Jacques talked to Madeleine as if tomorrow was promised. He remembered the promise that he had made to his wife during their wedding reception. Jacques told her that he could not wait to have their second wedding ceremony; Madeleine let her husband talk while remembering what he had once promised her the day of their ceremony. Jacques reminded his wife that she could have anything that her heart desired, hoping it would give her another reason to hold on for a little while longer in hopes of a donor match. Madeleine, at that moment, laid her head back while being cradled in his arm while listening to his soothing voice! The only thing Madeleine managed to say was, "Yes, honey, it does sound nice!" Madeleine closed her eyes as they filled with tears, as they began flowing backwards into her ears. Madeleine knew without Jacques ever telling her that her days of life were drawing near.

One morning, Jacques went to take the children to school, and once he arrived back home, to his surprise, he found that Madeleine was awake; she called Jacques to come to her side. Upon entering the bedroom, Jacques got down to his knees at the side of the bed.

Jacques placed his head on his wife's lap while she was lying. She said, "Honey, I am the luckiest woman in the world to have a love like you! Jacques, I want to thank you for loving me, supporting me, believing in me, and being the loving father that you are! I know that this is very hard for you, and I also know that this is very hard for our children," as she began to reminisce-about their days on the campus. Madeleine began talking about where they both loved to take many of their romantic walks and reminded Jacques of how wonderful his proposal was as they walked alongside their favorite spot along the water's edge! Jacques right then could no longer hold up in front of Madeleine, and she softly caressed Jacques's head as he sobbed from his heartfelt pain. She compassionately said, "Jacques, I need you to keep living for me!" What Jacques did not know was that Madeleine had overheard him praying the night that he was begging to trade places with her because she was awake that night listening to her husband's prayers for her! Madeleine said, "Jacques, I need you to please continue to make life great for our children," but he could not even speak for a moment. Jacques questioned Madeleine ever so softly while choking up, as he asked her, "How am I supposed to live without you; tell me how?" It was at that precise moment that Madeleine Bouvier succumbed to her diagnosis! Madeleine died at the early age of 37-year old. She had left all of those whom she'd loved, leaving them shattered and desolate!

Chapter 03: Without You

Jacques was devastated by his wife's passing. His only recourse for his grief was to bury himself in his law cases, trying to ease his pain. When Jacques was not winning cases, he was at home making sure that he was caring for his children.

Jacques had become so consumed with grief that he barely understood that his children were just as lost without their mother. The baby boy Bash turned three years of age just before his mother's passing! Bash clung to Gia as he had never done before, looking for comfort as he had always done with his mother.

Gia also clung to Bash because she needed to feel that same nurturing love that only their mother had given to each one of them. Every time Bash climbed into Gia's lap, she hugged him snuggly, just as she had seen her mother do with him. Gia would close her eyes and then lay her head on top of Bash's head while remembering her mother's comforting touch.

One evening, Gia put on some of her mother's favorite perfume right before bedtime. She knew that Bash would come to crawl into her lap before bed. Once again, Gia closed her eyes and remembered her mother's scent as she held her baby brother. Gia was certain that Bash was also remembering their mother because he had become very still.

Gia noticed Bash taking in slow breaths while rubbing the ends of her hair, much like he had always done with his mother because he could smell his mother's sweet scent as he and Gia both drifted off into a peaceful sleep. This was something that Gia and Bash consistently did for the next year, but they eventually weaned themselves off.

Jacques was all alone in Chicago without any assistance from either of the families because they lived back in New Orleans, and no one wanted to make a move to Chicago. Jacques was doing the best he could with all that he knew to be with his children; their lives without Madeleine were taking a toll on them all! Life felt like unchartered, strong-moving waters since the death of Madeleine, and Jacques seemed content when he was alone with his children, but his broken heart was taking a tremendous beating.

Jacques was seeing many parts of Madeleine in each one of his children! Jacques mostly saw Madeleine in his daughter Gia. He had noticed it more out of her due to her outward behavior. Gia was trying to keep things the same as if Madeleine had never died.

Jacques thought that Gia was for sure in denial because he had seen that she was looking for her mother, even in a perfume bottle, but never realized that she was doing it more for Bash so he could remember something about his mother.

Gia was doing as she had always seen her mother do daily. She made sure that everyone was up and always on time. Gia made sure that everyone was eating right and that everyone was exercising, also making sure that everyone was getting their homework completed. Even though Gia was only the third from the youngest child, she stood up for most of the responsibilities.

Gia was taking on the mothering role of her baby brother Bash; she did as much as she could do with all that she had seen her mother be. The older boys immersed themselves in playing sports or at the library, trying not to think of the loss of their mother; however, Angela always stayed in her bedroom, either reading or watching television when she was not at school. But that soon took a drastic change in her shortly after her mother's death.

The teachers had noticed that Angela was not coping very well since the death of her mother, which they expected, but it seemed she was not getting any better! The school decided that they needed to contact Jacques, and after the phone call, he rushed right to the school! Once he arrived to speak with the Administrator regarding his daughter, the Administrator told Jacques that Angela cries most of the time when she is at school, staring into space.

The Administrator said, "Mr. Bouvier, I know it is hard losing your wife and your children losing their mother, but I believe your daughter and perhaps the rest of your family, Mr. Bouvier, could benefit from grief counseling."

Jacques humbly said, "Thank you very much for your concern, but between my law cases and taking care of my children, I'm managing quite well!" Jacques's eyes started to swell tearfully, and right before one tear could drop, the Administrator said, "Mr. Bouvier, I lost my wife a couple of years ago. I know how rough it is with what you and your family are going through! I want you to take my number. Please feel free to call me anytime. Besides that, I would be honored just to listen if you ever need someone to talk with! Sir, what I can say is just take it day by day, and at times, it can even be minute by minute!" Jacques listened to what the Administrator had suggested, and once he arrived home before the children went to bed for the night, Jacques began trying to talk with Angela, but all she did was cry and was unable to communicate with her father!

Jacques realized that he needed to act fast. The next morning, he went straight to his phone to locate a grief counselor and then made an appointment for his daughter. The Counselor made a same-day appointment based on what she had been told by Jacques. Once they arrived at her office and after talking with Angela, she immediately identified separation symptoms and that Angela seemed very unattached to reality. The Counselor realized that Angela was giving her and her father answers that she thought they wanted to hear, but the Counselor then had her wait in another room before speaking any further with her father.

The Counselor explained to Jacques in more detail that Angela was having a very difficult time dealing with the loss of her mother; however, she believed that with continued treatments, Angela could return to her normal mind state, but it was not going to be a fast process! What the counselor made sure of was to express that it was going to take some time for her young mind to readjust and that she had a long road ahead of her! Jacques quickly said to the Counselor, "However much time it takes is fine. I just need my daughter to survive, and I need my daughter to be mentally healthy once more because my family has suffered enough with the loss of my wife and their mother."

The Counselor asked Jacques, "How are you coping with your grief?" Jacques smiled as the tears began filling his eyes. He felt as if the Counselor was peering into his soul while watching his breaking heart! Jacques tried to speak, but his voice began breaking. With a wavering tone, he said, "I lost my wife, my friend, my confidant, my lover, and the mother of my children, but to honestly answer that question, all I can say is that I am still here," as he gasped for air. The tears slowly rolled down from his eyes to his cheeks; however, the Counselor could see that not only does Angela need help, but so does Jacques.

The Counselor then stated, "I will see you both next week, Mr. Bouvier," not wanting to stress him out any further than what she had just witnessed! Jacques humbly reached into his suit jacket pocket and then took out a cloth handkerchief. He wiped his face before allowing Angela back into the patient's study room because he didn't want to let her see his tear-filled face!

Jacques did not realize that every time he took Angela to her appointments, the Counselor was leaving time on her appointment calendar to talk with him; she realized that he was a man in despair and that she needed to continue trying to reach him concerning his grief issues.

The Counselor was only into a few months of counseling with Angela, and while sitting at her desk one afternoon, she just happened to glance down at her scheduler when she noticed that Angela was late for her appointment. This greatly concerned her because Jacques had never been late nor canceled any of her appointments before.

The Counselor became concerned, and that was when she decided to call Jacques's office to see if everything was alright or if he had gotten stuck in court. The Counselor was able to reach Jacques's paralegal, Michelle, and she asked, "May I speak with Jacques Bouvier?"

Michelle sounded as if she did not know exactly what to say. She asked who was calling, hoping that it was not one of the children or their schools trying to reach Jacques because she was not prepared to explain what was going on at that time. The other Partners were all over the place at that time, just as confused as Michelle.

The Counselor stated, "I am with grief counseling for Mr. Bouvier, who was supposed to be here with his daughter this afternoon." She then went on to say, "And it isn't like Mr. Bouvier to miss any of his daughter's appointments!" Michelle said, "I can tell you that Mr. Bouvier did not forget, and he will not be making any more appointments any longer." The Counselor said, "I don't understand! What do you mean he will not be making any more appointments for his daughter because he understands the seriousness of these treatments?"

Michelle said, "Ma'am, there is no other way for me to tell you this, but Mr. Bouvier had a massive heart attack and died in his office this afternoon!" The Counselor's voice fell silent from shock for a few minutes. She then asked Michelle, "Had he seemed ill?" Michelle said, "No, he was healthy as far as I knew, but I heard his partners say that they believed it was due to stress from overindulging himself in cases and from raising his children alone." Michelle said, "In my opinion, I think it was mostly due to a broken heart after losing his wife. He had not been the same since her death because she was the light in his life."

Jacques Partners immediately made sure that the children were sought after until the family was able to arrive for the children. The Bouvier children were taken and moved back to New Orleans. They were raised by both Jacques and Madeleine's families until they were old enough to take care of themselves.

Chapter 04: Raise Up

The oldest of the Bouvier children, Jacques Jr., right after graduating from high school, decided that he wanted to join the military so he could start his career in law from the free education that the military offered. Andre, the second son, a year later followed in his brother's footsteps, joining the military with an interest in education. The third son, Kingston, also followed suit, joining the military, but he was undecided on a career path for himself.

Angela, the fifth oldest child, had continued signs of mental sickness. It was as if her mind could not progress any further growth. The actuality of it all was due to the combination that came from the loss of both her parents. The tragedies had pushed her over the mental breaking point of no return.

Gia, the fourth oldest child, was still mothering and raising her sibling Bash; her plate had become more than full. Gia seemed well-equipped by what her parents had taught her. However, the mothering of Angela with her mental issues was taking a toll on her. The one thing Gia dared not to do was leave Bash without a mother figure, she understood that she and their mother, Madeleine, were all that Bash knew.

The years rolled by quickly since the death of the Bouvier's parents. Angela was now 18 years age and still hadn't progressed mentally. Gia was now 19 years of age, and she was realizing that Angela's mental sickness was more than she could handle. She was going to have to decide to institutionalize her sister because her mind had come to a complete stop and was mimicking that of an eight or nine-year-old child.

The doctors explained to Gia that it was best to institutionalize her sister because Angela was incapable of handling any more delicate situations that could mentally take her out of being able to communicate at all. Gia was feeling the pressure from having to institutionalize Angela; however, she knew she needed to do what was best for his sister.

Gia clung even closer to her younger brother, Bash. She understood that she did not want him to end up in the same situation as Angela. Gia continued taking care of Bash. Several years later, after having to institutionalize Angela, Bash had become of legal age, and he had begun to venture out more, making his own friends. Where he would be able to care for himself, Gia knew that Bash had probably forgotten their parents and that she had become the only mother figure he knew.

Bash also had started to lean away from some of the teachings of Gia, while trying to learn to become his own man.

Gia told Bash that they were going to move back to Chicago because she knew that their parents would want them to remain there. She also explained to him that there was a reason that their parents had originally left New Orleans to try to give them more opportunities.

The Bouvier's children were taught valuable lessons by their parents for them to make sure to seek out the best-educated qualities with which they were raised. Jacques and Madeleine told their children that they would understand how prosperity works by caring for one another and that it would prepare them for when they were to have their own separate families.

The Bouvier parents instilled in their children that there is nothing more precious than family and for them to be mindful of whom they settle down with!

They also taught them that family can't pick your mate for you, but make sure to listen to the family's advice and to take heed of what your family says because their interest and concern are only out of love for you.

Chapter 05: The Narcissist "Natalie Montgomery-Bouvier" And the Montgomery Family

Natalie's a 5 ft -6 inches slender, wide-hipped, 100-pound young woman. Natalie's complexion is the color of that light brown paper sack. She has small, beady eyes with a broad nose that spreads across her face.

Natalie's upper two front teeth have a slight gap, and one of the teeth has a slight chip at the inner part of the gap. Natalie is an average-looking woman, but what best describes her is her quick, sharp tongue and her sarcastic attitude. Natalie is always looking to get someone told or to run them down.

Natalie's father, Ray Montgomery, is a 6 feet tall, handsome man with a dark chocolate complexion. Ray has thick, dark, curly hair and is a master barber by trade. He learned his barbering techniques at an early age which were passed down to him from his father. Ray's a spiritually fearing man who also serves as a Deacon in the church. When Ray isn't cutting hair or serving up his neighborhood, you can find him at the church. Ray fell in love with his high school sweetheart. He had asked her to marry him before they graduated from high school, where they both first met when they had a class in cosmology together.

Natalie's Mother: Lila Downs is a 5 ft 6-inch-tall light brown skinned, curvy woman with wide hips and big legs. Lila wears a short, curly hairstyle that she keeps cut in a beautiful taper style. Lila Downs also loved singing in the church choir. Ray Montgomery and Lila Downs married when he was eighteen and she was seventeen.

They both knew that they wanted to open their own beauty and barber salon. It was Ray Montgomery's dream to own his salon alongside his wife. Ray and Lila spent as much of their time in the church as they did in their salon. Ray and Lila agreed that when they had children, they wanted them to learn about the "Most High" and also love him as much as they did and to serve in the church where they served.

In their marriage, Ray and Lila were blessed with six children. The Montgomerys had also moved their parents with them from Georgia to Chicago, where they would also have better opportunities with their career choices.

The Montgomerys became well-established in their community; after they had opened their very own barber and beauty salon. Ray and Lila had both grandparents living with them to help with raising the children as they grew their business.

Ray and Lila were also very active in the church and after the children were old enough to walk and talk, they too also served in the church, which at times seemed that they all lived there. If you could not find the Montgomery at home or at the salon, all you had to do was go by the church. Once Ray said to the children to grab their Bibles, the children knew whatever they were doing had to come to a screeching halt.

The oldest of their children is Lisa, who has already begun her dental career. Then there are Natalie and Adrian, who are only a year apart in their junior and senior years of high school.

Melanie, the middle girl, is in her freshman year. Ray Jr. and Roman were still in junior high.

The one thing Ray and Lila wanted to make sure of was that they were going to give undivided attention to each of their children so that they understood how much they were loved. The Montgomery also wanted their children to make the best of choices for their lives. Therefore, Ray and Lila lived their lives in such a positive way that it was an example for their children to follow.

Ray and Lila's motto was, if you were not ill, then there were no excuses "not to" go serve at the church, but if you were ill, that's when one of the grandparents remained back at home with you, and they would have bible study with you for that day.

Natalie, the second oldest daughter, always claimed to be sick with either a headache or stomachache, and once they were done with the bible study, Natalie always said that she was going to bed afterwards. Natalie always managed to slip out of the house as soon as she knew that her grandparents were out of her way and she wouldn't be heard sneaking out of the house.

Natalie sometimes managed to arrive back before her parents arrived back home from church, but at times, she messed up and didn't make it back in time. Natalie would rather take the punishment for not making it back home in time with ease rather than be sitting in church all day.

Ray and Lila couldn't understand how Natalie was so different and difficult compared to the rest of her siblings. Natalie was hardheaded and very defiant despite her upbringing. Natalie would rather be somewhere else doing just the opposite of what she was being taught. She also hated being made to sing in the choir.

Natalie had a beautiful voice for singing, but she would rather be somewhere else, finger-popping and shaking her behind, and was interested in the guys that meant her no good, especially the ones who could have cared less about what her family could have wanted for their daughter.

Chapter 06: Bash and Natalie Meet

One evening, Natalie had gone to a skating party with her sister, Lisa. Lisa's fiancé, Ronnie, he had invited Bash, his best friend and roommate. Bash was more like a brother to Ronnie, and this was where Bash and Natalie's romance began.

When Natalie was first introduced to Bash, she looked at him, and her first thought was how handsome he was, but her next thought was that he would not be interested in her. Natalie believed that he would go for a much more attractive woman.

After a night of skating together with good conversation, she could tell that he was more than just a handsome man. He was a man with great morals. Bash kept his attention on Natalie for the rest of the night. As they walked the Montgomery women to their car, Bash asked Natalie if he could see her again. She ran her tongue across her teeth and sucked them while looking at Lisa to see what her reaction was going to be.

Lisa smiled and said, "Bash, why don't you come with Ronnie for dinner tomorrow after church?"

Bash, without haste, said, "Why don't we just come and go to the church with you all tomorrow." Natalie looked at Bash as if he was something strange. After the first Sunday visit, every time the Montgomerys looked up, Bash was headed to visit Natalie at Mr. and Mrs. Montgomery's house. Ray and Lila loved the fact that Natalie finally had a friend who was interested in going to the church, therefore, he was welcomed to their home any time. Natalie and Bash consistently dated for about six months when the family realized that Bash adored Natalie.

Mr. and Mrs. Montgomery were very stern with their children. They did not allow their daughters to have very much freedom other than to be in church from what seemed like Sunday to Sunday and a skate night on Saturdays, along with a Sunday dinner with some close friends.

Natalie always hated being made to attend church. She figured that if she were to get married then she could get out of her parent's house, so she was going to work her plan on Bash and make it happen.

There were times when Ray and Lila had found out that Natalie, of all their daughters, would tell them that she was going one place but come to find out that she had gone elsewhere. Natalie's parents could not trust her because they knew that she was a liar.

Bash grew very fond of Natalie. He then decided that he wanted her to go with him to meet his sister, Gia. One evening, Bash took Natalie with him and then stopped by Gia's house for his sister to meet his new love interest.

Bash rang the doorbell then Gia answered, but once Natalie saw just how beautiful Gia was, she instantly became insecure! Natalie felt uncomfortable before Bash's sister had even spoken a word to her. Gia smiled when she saw her brother. She then humbly welcomed Bash and Natalie to come in for their visit.

Bash turned towards Natalie and said, "Gia, I want you to meet my girlfriend, Natalie!" Gia looked at her brother and then stated, "Well, it is a pleasure to meet you, Natalie." Natalie quickly formed an opinion towards Gia. She thought to herself, this bitch is a stuck-up, then tossed Gia an arrogant snarling cool "Hi!"

Gia excitedly told them again to please come in, waving her arm in a manner that showed she was glad to see them! Gia offered them to have a seat. She looked at her brother and then quickly stated, "I have been missing you, and why have you not been by here to see me sooner?"

Natalie made sure she turned her full attention towards Bash, glaring with anticipation to hear his response. Natalie looked like she was watching the gears turning in Bash's head as he was trying to think of what to say to his sister.

That was when Natalie took the spotlighting opportunity and spoke up without haste, saying, "Because he probably has been spending it with me!" Gia slightly turned her attention towards Natalie and said, "Listen, I don't mean any harm; however, I was speaking to my brother!"

Natalie couldn't stand the fact that Gia had just corrected her, nor did she like the fact that Gia was questioning Bash, sounding like she was his mother. Natalie sat back on the sofa aggressively, and then immediately, she did what she does best when she becomes annoyed with someone. Natalie sucked her teeth and folded her arms to let Gia see that she was annoyed with her questioning.

Gia, without missing a beat, went right back to questioning Bash right after she was done correcting Natalie. "Bash, as I was asking, are you taking good care of yourself? And please tell me you are eating right."

Once again, Natalie opened her mouth to speak for Bash, "I can tell you he is doing just fine," but by that time, Gia had about enough of Natalie's meddling sarcasm.

Gia turned her full attention towards Natalie, looking directly into her eyes. She said with every bit of vigor in her voice. She said, "Listen, Natalie, and that is your name, right? But, like I said a minute ago, let my brother speak for himself, if you didn't notice. he does have a mouth!

Bash looked wide-eyed because he had never heard that side of his sister's personality! Bash quickly said, "I am doing fine, Gia. Everything is going great!" Bash felt he needed to appease both the women in his life to keep the peace! Bash believed he felt the lioness's heat, but what he did not realize was that there was only one true lioness amongst the two of the women!

Gia looked at her brother, wanting to say badly to stop being simple, but she calmed herself then said, "Well, Bash, as long as you're doing well and you're happy, that is what's important, but just know that because you are grown does not mean that I don't wonder about how you are doing."

Gia, right away, asked Bash as calmly as she could, "Do you think that you can come back later? I really need to talk with you about some things," but not before Natalie's agitation once again got the best of her, causing her to roll her eyes up in her head.

Natalie sternly stated to Bash that she was ready to go! Bash suddenly made an excuse that he had some other things that he needed to go see about. He then told Gia that he would be back later in the week.

Once Bash and Natalie got into the car, Natalie could not wait to say, "Your sister acts so damn stuck-up with her shitty-ass attitude!" Bash said, "Natalie, she is not a stuck-up at all. Gia is actually a very sweet person and I promise you are going to love her once you get to know her."

Bash told Natalie, "Listen, it's just that you'll need to learn our relationship, then it'll make sense to you. Gia has reasons to be concerned, that's all." Natalie not once ever asked Bash what he meant by their relationship was something that she needed to learn because she already knew that Gia was his sister, and that was all she cared to know about her.

Natalie quickly moved past that conversation and then asked, "So, what do you think she wants to talk to you about?" Bash said, "Baby, I honestly don't know, but what I do know is that she's made a lot of sacrifices for me, and I'm thinking that she might need something from me for once, but I'll go back soon to find out."

Lisa's fiancé, Ronnie, and Bash were roommates. Ronnie works as a head boxing coach at a gym that Bash frequents often, which was how they met each other. While sparing with Ronnie one afternoon, he kept dropping hints for Bash regarding Natalie; however, he chose to ignore all the warnings.

Bash kept talking about Natalie and Ronnie and kept saying, "Man, you need to slow down with Natalie until you get to know her better." Ronnie also mentioned that he does know that her parents have had a difficult time with her and with some of her dating choices. He then said, "Listen, I even had to put my hands on one fool that she had been dealing with after he put his hands on her."

Ronnie chuckled and said, "I tried to tear the lining out of his ass! Now, I'm not saying that I would lay hands on you, but before Ronnie could say another word," Bash said in a joking manner, "You know I got these hands too as he punched at Ronnie!" Bash said, "I learned from a master!"

Ronnie said, "All I am saying is don't make me have to pull up because I'm skinning everybody's ass up about my brother!" Bash said Ronnie, it is all good, brother. Now go get your crazy behind back on that chain! Ronnie turned around and looked back at Bash, then gave him a deep growl and started laughing!

Gia kept calling for her brother to check on him because a week had already gone by, and he had yet to return to see what she wanted to discuss with him. When Bash was with "Natalie" and Gia called her brother, he kept telling Gia whenever he answered her calls that he would call her back but never did.

Gia was becoming very concerned for her brother because she did not care for what she had already seen out of Natalie! Bash finally stopped by Gia's late one evening before going to work at the post office.

Bash finally went to see what Gia needed to talk to him about. They went into the house and took a seat in the living room. Gia asked her brother, how come it took him that long to come back? Bash responded by telling Gia I have been extremely busy, and I've been spending a lot of time with Natalie.

Gia said, “Bash, that’s exactly what I wanted to talk to you about! I don’t know anything about her, but I really did not get a good impression of her when you both were here.” Bash swiftly came to Natalie’s defense, saying, “Look, Gia, you don’t even know her to be forming any opinions of her.”

Gia said, “You’re absolutely right. That is why I asked for you to come back so we could have this talk!” Gia told Bash, “What I seen was that she does not know you well enough, hell or me for that matter, to be interfering with me asking you anything.” Gia went on to remind Bash of what their parents had taught them. She said, “Bash, I know you don’t remember, but Mom and Dad both always told us to take care of family and for us to keep watch over each other. Now, I know that you are a grown-up, but I felt that Natalie might mean you no good! Bash, all I am asking is that you be careful when it comes to her.” Bash said, “I know you love and want the best for me, but just listen, sis, Natalie is good to me, and she is good for me!” Gia nervously said, “I just hope that you’re right, but please take caution when it comes to her!”

Chapter 07: The Perfect Victim

Natalie made up her mind that she was going to get pregnant so that she could get out of her parent's house sooner than later. Natalie had grown extremely tired of her parents' strictness and being forced to go to church 2 to 3 times a week.

Natalie always closely kept watch of how Bash longed for his family, each time after he visited her at her parent's home. It seemed like all he ever talked to her about was how much he missed his siblings and one day, he hoped to have a family as close to what he witnessed with hers.

Natalie figured Bash would marry her if she became pregnant. She then made an immediate decision to stop taking her birth control pills and began having sex with Bash until she was able to achieve her ultimate goal of getting out of her parent's house by any means.

Natalie began sneaking out of the house late at night, going over to Bash's apartment because she was more than adamant about getting pregnant, and it had become an intensely obsessed mission to make sure to see him each night.

Natalie finally accomplished missing her monthly menstrual cycle; however, she waited until she was two cycles late, making sure that there was no doubt that she was pregnant before revealing it to anyone. Natalie was about to explode with sneaky excitement to tell Bash that she was pregnant because she knew how happy he would be. Natalie knew that she could get Bash to do whatever she needed him to do because she was able to get him to distance himself from his sister and feel she had control over him.

Natalie planned on telling Bash that she was pregnant that coming Saturday night when they were to be going to the skating rink. When Natalie and Lisa stopped to pick up the guys, Natalie said, "I am not going skating tonight. I am staying in with Bash because we need to talk!" Lisa was wondering what it was that Natalie needed so desperately to talk to Bash about that she could not talk to him about it in the car or at the skating rink.

Lisa sarcastically said, "Look, Natalie, you better have your behind here and ready to go when I get back. Do you hear me?" As Natalie turned to get out of the car, she sucked her teeth, twisted her lips up, and rolled her eyes up in her head. Natalie turned halfway around and said, "Girl, whatever," then flicked her sister off with a dismissive wave of the hand.

Bash was heading towards the car with his skates over his shoulder when Natalie sternly said to him, I need to talk to you! Bash looked confused because he didn't know why Natalie would be sounding ever so more direct with him. Bash had no idea what she needed to talk to him about, so when they went into the apartment, Natalie took a seat on the couch with her head hanging low.

Bash took a seat next to her, looking attentively at Natalie's face. She started to become emotional as tears began to swell in her beady eyes. Natalie's face turned red, and he did not know what was wrong or what to do. Bash asked her, "What is it? What is wrong that has you this upset?"

Natalie boldly looked Bash in the face, then hysterically blurted out, “I’m pregnant,” but the pressure game was just getting started. She immediately said, “You are going to have to marry me!” The antics were well-orchestrated as she turned, then began to sob all over his shoulder.

Natalie rushed to explain, stating that she did not know what had happened and she was taking her birth control pills, but her cycle had not come for two months. Natalie said, “Bash, all I know is that my parents are going to kill me. I don’t know what I am going to do besides that. I know that my parents are going to expect us to get married. I am just so scared that I don’t know what I’m going to do!”

Bash sat there for a minute, taking in what Natalie had just told him, but before she could say anything else, Bash said, “Natalie, I love you very much, and of course, I am going to marry you! I had already planned on asking you to marry me sometime in the near future, but I suppose now is that time, so calm down, please!”

“Natalie, you don’t need to worry. I will be with you when we tell your parents our news. I know that they are not going to kill you; now, I just hope that they don’t kill me first! Natalie, you know that family is the most important thing to me. I guess we will need to speak with your parents and tell them tomorrow before I go to work.”

Natalie continued with her antics of crying as Bash wiped her tears. Bash then said to her, “It will be alright besides that. This is good, right?” Natalie looked up at him with her red face and puffy eyes, still crying, then responded by saying, “I guess so!” Bash was already thinking ahead, realizing that they were going to need a larger living space. He then said,

"Well, I will go to my sister to see if we can stay at the house with her or at least until after the baby arrives. I'm sure that it should not be a problem with Gia, it will allow us time to be able to save up to get a bigger place."

Now, this is the one thing that Natalie had not seen coming from her well-put-together scheme. She did not expect Bash to bring Gia into her plot; however, she knew she had to keep her mouth shut because she did not want to cause any waves until after Bash had married her.

Lisa and Ronnie had arrived back from skating to pick up Natalie. By then, she had pulled herself together and was back to her usual acting self. Lisa asked Natalie once she had gotten into the car, "What was so important that you all had to stay in tonight to talk about?"

Natalie sheepishly looked out the passenger window with a shit-eating grin on her face, then said, "You'll find out soon enough! Lisa asked, "What are you talking about? I really hope that it does not involve Bash because he loves you!" Natalie sarcastically said, "I know that he does, and so now, what are you, his keeper?"

Lisa asked her, "Please tell me you didn't break up with him because Ronnie mentioned to me that he does not think that you really love Bash in the way that he loves you; please tell me the truth, how do you really feel about Bash?"

Natalie said with an annoyingly quick snare, "He'll do!" Lisa said, "He is a great guy, and I pray you don't mess it up." Natalie responded by saying, "Here you go with your holy roller bullshit, but you should be more worried about you and Ronnie's behinds. However, just make sure that both of you stay the hell out of our business!"

Bash and Natalie went to speak with Mr. and Mrs. Montgomery the following day to inform them that they were expecting a baby. Bash also wanted to ask her parents for their blessings to marry their daughter; however, Natalie just sat there, quiet as a deaf-mute individual. Once her parents came in, they sat down to listen to what they needed to speak with them about.

Natalie was letting Bash do all the speaking and never once intervened to let her parents know any plans regarding Bash and herself. Natalie's parents were shocked and disappointed that she had gotten pregnant before she had gotten married, but once they heard Bash explain to them just how much he loves Natalie and that he had been planning on asking her to marry him before the pregnancy!

Ray and Lila Montgomery could not help but give them their blessings. They then welcomed Bash into their family. Bash took Natalie along with her parents the following week to the city courthouse, where they had a quick marriage ceremony in front of a Judge. After the ceremony, they all arrived back at her parent's house for a reception with most of their family and a few of the church members. Bash did not invite Gia or tell her about the news because of the tension between his sister and Natalie but he was going to let her know soon.

Bash and Natalie stayed at his and Ronnie's place for a few days until one afternoon. Bash told his new bride that he needed to go to speak with Gia about them staying with her until the baby arrived. Natalie quickly said, “Why don’t we just stay here?” Bash said, “Because her home is larger, and it’s better suited for when the baby arrives. Besides that, I don’t want to impose any further on Ronnie. Natalie, it’s about to be three and a half people living in this small, cramped apartment, and I won’t do that to Ronnie.”

A few days after the wedding, Bash decided to go alone to visit Gia to share the news. He was somewhat a little nervous because of the last conversation that he and his sister had regarding Natalie. Once Bash arrived and Gia answered her door, she was excited to see that there stood Bash, but he was looking more like the cat who had just swallowed the canary when she looked at him!

Gia excitedly asked, “How are you, and what do I owe to this great surprise?” That was when Bash said, “Better than I expected.” Gia said, “That’s great, but what does that mean?” Bash said, “I have some awesome news I want to share with you!” Gia said, “Well, come on in here and tell me all about your awesome news.”

Bash said, “Natalie and I got married a few days ago!” Gia’s mouth flung open, but nothing was coming out. Gia finally said, “Bash, stop kidding.” He gave his sister a slight smile while looking directly into her eyes and said, “I’m not kidding., Gia. Seriously, we got married a few days ago, but hold on to your top and Natalie is also pregnant!”

Gia was stunned. She softly asked, “What did you just say? Bash, Wait, did you just say that Natalie’s pregnant? When is the baby due?” He said, “In about seven months.” She asked, “Was the baby planned?” He said, "No, we hadn't planned it, and she was taking the pills, but the fact remains that she is pregnant!"

Bash went on telling Gia, "She comes from a family that does not believe in children out of wedlock, and I was not going to let her birth my child into this life without the Bouvier name. Besides that, I do love her! I was planning to ask her to marry me anyway, but things happened sooner than I had planned."

"What I need to do now is to find us a larger place to live before the baby arrives." Gia instantly extended an olive branch to her brother by asking if there was anything that either of them needed that she might be able to help them with. Bash said, "As a matter of fact, we really need a place to live because the apartment is too small, and I was hoping we could stay here with you until I save up more money for a larger apartment."

Gia said, "Of course, whatever I can do to assist, besides that, I will get to be close to the first Bouvier grandchild." Gia went on to say, "Just make sure you both use this time to get on your feet, and there won't be any rush on either of you while you are here."

Gia ended the conversation by saying, "I am so very proud of you," which made her brother feel even happier, knowing that his sister was pleased with his marriage and the excitement of the baby that was coming.

Chapter 08: Light of Truth

A week later Natalie and Bash packed up their belongings from the apartment and then showed up to move in with Gia. The moment Gia approached Natalie, she excitedly said, "Congratulations, Natalie, on the marriage and the baby!"

It was right there at that moment that Gia was greeted with a short snub by Natalie, she responded to her with an uninterested thanks and caring not to elaborate any further with her. Gia had seen her haughty demeanor but thought that it could be from maybe having morning sickness or being tired from moving; therefore, she did not take offense to it.

Gia walked into the house with them both and said, "I've already moved out of the master bedroom, so you all can take it because you all will need it for when the baby comes home." Each day thereafter, when Bash left for work, Natalie never came out of the bedroom to engage with Gia other than to grab something quick to eat or to use the restroom, and then, she would retreat right back to their bedroom.

Gia periodically would go and knock on the bedroom door to see if Natalie needed or wanted anything to eat after she had cooked, but Natalie always refused any food offers from Gia.

Natalie kept a supply of snacks in the bedroom, so she didn't need to eat any of Gia's cooking. Natalie kept consistently complaining to Bash about Gia, stating that she did not prepare any of the foods that she cared to eat. One afternoon, Natalie told Bash, "No wonder your sister's ass is so damn skinny. It's because she does not eat anything but rabbit food."

Natalie angrily told her husband one evening before he left to go to work. She said, "Listen, you need to bring some real food in here because I am tired of your sister telling me how I should be eating." Each night, Bash had to stop on his way home from work just to bring Natalie the food of her choice, either from her parent's home or from a restaurant. Natalie didn't understand why Gia had always eaten a healthy diet and walked for exercise out of fear that she didn't want to end up like her parents, who both died at such early ages.

Natalie also didn't like that Gia never brought junk food into the house, which was something that she was not accustomed to. Natalie stayed on the phone with one of her sisters, forever running Gia down when Bash wasn't present, which had become a part of her daily ritual. One afternoon, before Natalie had gotten on the phone with her sister, Adrian, who was the one she always entertained her foolishness to, Gia knocked on Natalie's bedroom door right before taking in her afternoon walk, but there was no response from her. Gia softly said, "Hey, I am going out for my walk." She assumed that Natalie must have been asleep because there wasn't any response.

Natalie heard Gia talking to her, but she chose not to answer! Once Natalie heard the front door open and then close, Natalie assumed that Gia had left. Natalie went and picked up the phone, then called her sister Adrian to bear her daily frustrations concerning Gia. As soon as the call connected, Natalie said to her sister, "Girl, let me tell you something about this skinny heifer; her sorry ass can't cook worth a shit." Adrian had already begun laughing once the call connected!

Natalie said, "Why is she always cooking them damn bland mashed potatoes and those nasty-ass sugarless pies?" Natalie went on, saying, "No, but wait, it gets worse than that! Adrian. Listen, last night, she got fancy with it and prepared alongside them sorry, lumpy potatoes, some green beans that she drizzled with some sort of vinaigrette sauce. Seriously, what the hell was that? Her damn ribs should be sticking to her back."

Adrian damn near couldn't even breathe from laughing at Natalie; however, she managed to ask Natalie, "Where is Gia while you are talking so bad about her?" Natalie said, "She took her boney ass out for a walk as if the bitch needs to be walking!"

Adrian again was out of breath from laughing but said, "You all need to just come to stay at Mama's and Daddy's house until the baby arrives if it's really that bad!" Natalie had a rushing response to her sister's comment. She loudly said, "Hell, that's exactly why I got pregnant in the first place, just so I could get the hell out of them tambourine-banging folk's house!"

Adrian's shocking response was, "What?" Natalie realized what she had just said, but before her sister could ask anything else, Natalie asked, "Why would I come back there?" There was about to be a surprise for Natalie. What she did not know was that Gia had never left because it had begun to rain once she had opened the front door to partake in her afternoon walk.

Gia stood at the front door for a minute, then said to herself, "Well, I'll have to try later on to get my walk in." She crept back down the hallway into her bedroom, trying to keep quiet and not wanting to disrupt Natalie's nap, or so she thought.

Gia had returned just in the nick of time. She heard the entire conversation between Natalie and Adrian in its entirety due to the call being on speaker phone. She heard everything that Natalie had said, and she swiftly but silently moved from her bedroom in a furious rage! Gia stormed to where Natalie was then forcefully pushed open the bedroom door, and that's when Natalie saw that Gia was damn near breathing fire which startled her!

Gia harshly said, "You are about one unappreciative trifling hussy!" Natalie looked like she had seen a ghost because it showed on her terrified face. She didn't know if Gia was about to attack her as her mouth dropped open.

Gia angrily said, "Bitch, close your mouth. I knew your raggedy behind was up to no good, Natalie! You are one lowdown snake in the grass. I tried to warn my brother about you from the first time that I, unfortunately, met you. Let me educate your cheap ass about something. My cooking is what keeps me healthy. Just maybe if your wide ass were to give it a try, you might not be walking around here looking like your behind is wider than some damn bull moose!"

"Natalie, apparently, your ignorant behind can't know that our mother and father died when Bash was only three years of age, and oh yeah! For your information, since I already know what you stated to Bash that you don't like the fact that I act like I'm his mother! Now, before I get into that, after hearing the conversation between you and your sister, I also learned that you're also a comedian! Please entertain me and let me ask you a question. Natalie, who in the fuck do you think raised Bash?"

Natalie had no clue. She remained silent! Gia instantly took the opportunity and said, just as I thought! Gia then let loose what sounded like some sort of morbid, sinister-sounding chuckle while looking directly at Natalie!

Natalie's behind was for sure that she was about to not make it out and was afraid! Gia said, "I'm saying this. I understand that you would have to sincerely care to know that Bash sure didn't raise himself!"

Gia was trying to control her breathing by taking in slow and deep breaths while praying not to lay hands upon Natalie! Gia also said, "Let me inform you of something. I am his mother figure, you trick-ass bitch rather you like it or not, so deal with it!"

Natalie was thinking about making a run for it, but she knew that Gia was in shape and that she would probably catch her the moment she made any sudden movements! Natalie thought to herself, this boney bitch is stealth-like because I didn't even hear her when she came back in, so I should just stand still!

Natalie had forgotten for a minute that her sister was still holding onto the phone, and with all guilt in her voice, the only thing Natalie was able to say to Adrian, who was still listening, was a low, heavy whisper as she said, "I'll call you back!" Once Natalie hung up the phone, Gia continued tearing into her behind!

Gia said, "I told Bash that he needed to watch your tainted behind. I instantly recognized that you have a mental problem along with some sort of fucked up dark heart! I think, nope, scratch that shit; I know it is best that you get whatever your raggedy ass came in here with and get the hell out of my house!"

"Natalie, you need to know that I don't give a rat's ass if you move back into your parent's house. What I can't begin to fathom is that your parents could possibly know just how low down you are! Gia said, "Now I'm about to leave because I don't want my son to see just how upset I am, but mainly for the reasons why!"

Gia made sure not to be home when Bash was to arrive back in from work. She did not want to have to tell Bash the truth as to why she was angry with his wife. Gia knew it would break his heart to know that kind of sadistic behavior came from the woman that he had just married and with whom he was about to have a baby with.

Natalie had nothing to say during the entire altercation. The only thing that she was wondering was if Gia was going to tell Bash about the conversation that she had overheard between her and her sister.

Natalie was glad when Gia finally left the house because it was going to give her time to think of an excuse for how she was going to convince Bash that she was homesick and that she needed to go home to her parent's house that night!

Once Bash arrived back from work, he saw that Natalie had already packed up their belongings! Bash looked at his wife, completely confused, and asked, "What is all this?" Natalie said, "Bash, I need to go home because I am missing my parents. I am not used to an environment where I'm not with my family."

Natalie began her antics of crying and said, "I just want to go home right now, Bash!" Bash said, "Listen, I am tired. I've been working all day, and this will just have to wait until morning!" Bash went on to say, "Gia just moved all this furniture around so that we could move in here! Please tell me what we're supposed to tell her, Natalie. This doesn't make any sense for you to be acting like this because we are married now and with a baby on the way. What I don't understand is that you were ok while we were at the apartment with Ronnie. Listen, we can't just be jumping up and running to your parent's house because you are homesick!"

Natalie said, "Well, either you can come with me or I will be going by myself!" Natalie was certain that Bash would follow her, but she needed to forcefully secure his decision by adding insult to injury.

Natalie said, "Bash, I can tell that Gia does not like me, nor does she really want us here." Bash asked Natalie, "Why would you say something like that?" Natalie said, "Because if she did, she would act a lot nicer to me. Besides that, Gia knows that I am pregnant, and there would be food in here other than that rabbit ass shit that she keeps cooking."

Bash said, "Natalie, there is food in here. My sister just eats a healthy diet which you've seen how I have also eaten as well, so I don't understand what the problem is so suddenly?" Bash said, "You do realize that you can go in that kitchen and prepare whatever foods that you like for yourself, right? I'm pretty sure Gia would not have a problem with you preparing and eating whatever you like!"

Natalie's only response was, "Well, at least at my parent's house, I can get some real food!" Bash looked at Natalie and shook his head with disbelief. He was fed up with her nonsense. Bash said, "You know what, Natalie?" while fighting, trying to suppress his tongue. He then pondered for a moment. Bash immediately took in a deep breath and then said, "Just never mind." He headed towards the bedroom and began packing up the car. He drove quietly, taking Natalie and her belongings to her parents' house.

Bash went in first to explain to Mr. and Mrs. Montgomery that Natalie had become homesick and was crying to come home. Bash confided in Mr. Montgomery, saying, "I don't know what is wrong with Natalie or why she is acting childish like this!" Mrs. Montgomery heard what Bash said to her husband, but she wasn't buying her daughter's act. That's when she looked at Natalie with this questionable stare on her face!

Mrs. Montgomery was wondering why, all of a sudden, Natalie had become homesick when she was always sneaking to get out of the house. Mrs. Montgomery knew that there had to be something more to the story, knowing her daughter's previous behaviors. Natalie would not even look at her mother. She was hoping that her mother would not begin to question her in front of Bash.

Bash asked if they could stay there with them until he could find a place closer to their home. Mr. and Mrs. Montgomery told Bash that it was fine for him and Natalie to stay there for however long it was going to take. He thanked them and said, “Hopefully, it won’t take too long, and I’m hoping to find something before the baby arrives.”

Bash made sure before they left to leave Gia a note explaining why they had left so abruptly! Natalie was already in the car, ready to leave, and did not know that Bash had left the note for his sister. The note read, “I am very sorry to have unexpectedly left like this, but I had to take Natalie home to her parent’s home because she had become very homesick to the point of hysterics, and I did not want to put any stress on the baby!” Bash also said that they would be staying with her parents until he could get them a place, but as soon as he found a place for them to live, he would let her know.

When Gia arrived back home and saw that they had moved out, she was saddened, but only for her brother! Gia found and finished reading the note her brother left, and it infuriated her all over again. Gia spoke out loud due to her frustration saying, “This harlot is going to make me hurt her lying ass!”

The next day, Bash found a nearby apartment close to Natalie’s parents’ house, which they could move into right away, and she could go to her parents’ house daily if she chose. Bash arrived back to let Natalie and her parents know that he had found an apartment. Mrs. Montgomery immediately took notice of how Natalie seemed to have instantly recovered from her state of homesickness and was eagerly ready to pack and move.

That's when Mrs. Montgomery said, "How interesting, Natalie, isn't it? What a full 24-hour period can do with homesickness, wouldn't you agree?" Natalie would not even look up at her mother because her mother said it while Bash was standing nearby.

Times like that, Bash seemed to be really missing his family, and rarely did the older Bouvier brothers get a chance to see Bash, nor had they gotten the opportunity to meet Natalie. Gia had completely distanced herself away from Bash and Natalie, not because she wanted to but because she had to.

Gia was constantly thinking about the last time when she and Natalie had an altercation with each other, however, she hoped that they could restore some sort of relationship before the baby was to arrive. Bash decided that he needed to call his sister one evening just to talk with her. He realized that it had been a while since he had seen or spoken with her! One day, Bash called and told his sister that he and his wife had found a new place, then offered her the new address. Everything seemed to be going well, but the obvious was noticed in how quiet Gia became each time Bash kept mentioning Natalie. He also noticed that his sister never once asked how Natalie was doing regarding the pregnancy!

Bash then kept going on about how homesick his wife had become the night that they had suddenly left, but Gia could not hold her frustration in any longer. Gia sternly said, "Bash, wake your stupid ass up because that bitch has your fucking balls in her hands!" Bash was shocked by her comment. He immediately jumped to defend his wife by saying, "Gia, that is my wife, whether you like it or not. However, you will not disrespect her."

Bash then went on to say to his sister, "You are wrong about my wife, and besides that, Natalie told me that she felt you did not like her!" Gia removed the phone from her ear as she glared at it for a minute in complete disbelief and wondered how it was that her brother could be that damn foolish.

Gia gave a sarcastic, indignant chuckle while trying to keep her anger under control. She said, "Yeah, ok, Bash, you believe what you like while trying to hold her tongue!" Gia thought for a minute, then said, "No, wait a minute, hell Bash, she does not even respect you, so how in the hell can you demand that I'm going to respect her?"

Bash asked, "What are you implying? That she doesn't respect me?" Gia fell silent, and Bash didn't realize that it was only out of love for him that his sister decided to keep her mouth shut about what she had heard from his wife. Gia did not want to lose her brother over Natalie's wicked ways. She finally said, "Bash, you just need to pay closer attention. That is all I have to say about your wife, and I'm done debating anything about her!"

Bash's brothers decided that they needed to call him after they had spoken with Gia later that same afternoon. The older Bouvier trio brothers made the call to their younger brother by a conference call to catch up on the latest and to see how he and Natalie were managing. Bash was excited to hear from his brothers. He was telling them how happy he was being married and that he could not wait for the arrival of their baby. As soon as the conversation was over with his brothers, Bash was immediately met with negativity from his wife. Natalie was listening from the other room and realized that it was Bash's family on the phone with him.

Chapter 09: Divide and Conquer

Natalie wanted to ward off all the "Bouvier" siblings before they could get into Bash's ears regarding anything about her and for her to maintain control over Bash. Natalie also understood that just because Gia had not said anything to Bash about what she had heard, she also knew that there was one thing for certain that Bash's sister had told or was going to tell her brothers or any other family member the truth about what she knew about her.

Natalie was glad that her family all lived in the same city and they spoke to each other every day, therefore it was easy for her to keep her lies well organized. Bash loved Natalie's family, and they loved him just the same, but the main reason the Montgomerys loved Bash was that they were certain that he wholeheartedly loved Natalie.

The Montgomery family never turned down any invites from Natalie and Bash at their new place, whether it was for dinners or some sort of party. The invites had become a weekly weekend event because Bash was a great host who loved being in a family environment, and he made sure that all needs and wants were in place for their guest.

Gia called her brother one evening to see how he was doing. She asked Bash if he would come by so that she could see him. Again Natalie was sitting in another room with her ears peeled back, listening to the conversation, and once Bash got off the phone, Natalie walked into where he was and asked in a hostile tone, "What did that bitch want?" Bash responded with a draining tone. He said, "Natalie, she was only calling to see how we are doing, that's all. Damn!"

Natalie hatefully said, "That bitch does not need to know nothing about me." Bash had grown sick of his wife's nasty attitude towards his sister. Bash asked his wife, "Why do you have to be so damn insulting by calling my sister out of her name?" That question pissed Natalie off even further, but she never gave him an answer regarding her dislike towards his sister. Once again, all she said was, "Like I said before, the bitch don't need to know shit about me!"

Bash knew right then that he was not going to mention that Gia had asked him to stop by for a visit just to keep hostility down to a minimum with his wife. Natalie was feeling the pressure from her exposed demise; however, she was right about one thing. Gia had talked with her older brothers, and she let them know the truth about Natalie and Bash, as what happened at her home the evening when she caught Natalie running her mouth to her sister Adrian on the phone.

Gia told her older brothers that Bash and Natalie had gotten married because she had slickly trapped him by getting pregnant in order to get out of her parent's house. She also told them that Natalie had been telling hateful lies about her to Bash and that Natalie did not like the fact that she carried the role of being Bash's mother. Gia then informed them of how she had to put Natalie out of her home but that she had not told Bash what really happened that evening because she did not want to be the bearer of bad news. The Bouvier brothers were also told that Natalie has her venomous fangs into their brother and that he no longer listens to any of the wisdom and concern for himself. Gia said she felt like Natalie was isolating him away from her.

Bash's brothers told Gia that they believed that Bash would soon come to realize that Natalie does not love him in the same way that he loves her and that he'll eventually stand up for himself. The oldest brother told Gia, "Just remember, he's a "Bouvier" and we were taught well, sis!" Gia said, "I hope you are right, Jacques Jr.!" Their sister told them that, at times, he does maneuver like their father in that "Jacques Bouvier Sr. ESQ" manner when it comes to love, but the difference was that their dad was a man who knew how to choose the right mate. That comment caused them all to lighten up on the original conversation because they were remembering their father's love for their mother.

Chapter 10: Betrayal and Hoodwinked

Bash and Natalie Bouvier's first child was finally born seven months after their marriage. Natalie gave birth to a baby boy whom they named Andre, after one of Bash's brothers. Natalie adored Andre, whom she placed upon a tall petal stool. She made sure that he wore the best of everything or nothing at all.

Bash and Natalie moved again, but this time, it was into a very nice brownstone right after the birth of their son. Natalie was still not happy with the new place that Bash had gotten for them. Natalie kept thinking about how Lisa and Ronnie had already purchased a new home right before they were about to get married and Natalie figured that she and Bash needed to have the same thing or something even better.

Once Andre started walking, Natalie decided that she was going to get a job. Natalie told Bash that she was ready to go to work so that they could get started on the purchase of a new home. Bash agreed with his wife that they should work together to purchase a home now that they have begun a family.

Bash switched his shift and worked the graveyard shift so he could be home with Andre while Natalie focused on finding employment. They also agreed that she should be working between the hours of 1p and no later than 9p. That way, their son would always have both his parents with him throughout the day, which would also save them childcare costs.

Natalie finally landed employment where she had taken on the position of a housekeeper at one of the local hospitals. She would be working with about eight other people and five out of those eight people were men. There was this man who worked in the housekeeping department named "John Harris." John's a 5 foot 6 inch chunky built man, who is light-skinned with sandy brown colored hair. John was also known as the mouth of the housekeeping department as he was one fast-talking, heavy-drinking man and was known for sleeping with most of the women who had already come in or gone out of the housekeeping department.

Natalie began her new position, where she seemed to always be watching and listening to John talking crazy as he kept her laughing. One night after their shift, a few of her co-workers decided they were going to stop at a tavern near the hospital that they referred to as their watering hole. John went to Natalie before the shift ended and asked her if she was going to join them for a drink, and unbelievably to John's surprise, it did not take much coaching for him to engage her into going with them all.

Everyone knew that John was known to be one slick, swift shit talking man who always kept one thing on his mind, which happened to be how much pussy he could rack up. Natalie was intrigued and mesmerized by his charismatic slithery charm, and she believed that she was slicker than his games; however, what she did not know was that she was an easy prey for John because he had already studied her for his plot from day one.

Natalie arrived at the tavern, switching her wide hips hard with hopes of catching John's attention. Natalie took a seat across the table from him. She kept looking around as if to not let her eyes fall upon him. John couldn't keep his lustful eyes off of Natalie. However, he had taken notice of how she was trying way too hard not to look at him, and he knew right then that he would have her soon.

Natalie had one drink before she decided to call it a night. She told everyone that she had to be getting home to her husband and son. John, with a wet-lipped side mouth grin, smiled and then said, "All right darling, we will see you tomorrow!" Meanwhile, John and the other men sat back as they watched her tantalizing switch walk towards the tavern door to leave.

John leaned back in his seat, then took the back of one of his hands with a slow effort and swiped across his forehead. He looked at the other men at the table, and he chuckled as if they had already read his mind. John told the men I think my notch might need another belt whipping. Natalie knew that John had his eyes on her while walking towards the door, she was hoping that no one else noticed her swinging her behind from side to side for John to watch!

John Harris kept his eyes on her behind, and he knew that he was going to get Natalie in his bed. John was the one who made up the housekeeping schedules for the staff. He instantly changed the schedules and made sure to set things up so that he and Natalie would team up during their shifts and he could work his plan. All the other co-workers knew how he groomed the women once he knew he stood a strong chance to sweep them into his grasp.

One evening on the job, when no one else was around, John boldly told Natalie, "I sure do like them wide hips that you be swinging around here." Natalie did not attempt to correct him on his sexual advancement but only said, "John, you would not know what to do with these hips if you had the opportunity." John laughed because she had just given him the green light to move forward, which was all that he needed. He then said, "You want to bet?" Natalie loved and craved his flirtatious response, so later that evening, before their shift was about to end, John asked Natalie if she would like to meet him for another cocktail and then stated that it was his treat.

Natalie said, "I am not fooling with you all this evening." That was when John said, "Who said anything about anyone else because I'm talking about me and you?" Natalie asked, "What do you want to talk about John Harris?" John confidently said, "You already know them wide hips that you keep swinging around here!" Natalie began blushing and laughing and said, "You are so full of shit." However, the conversation had her already hot and turned on by his advancement and his interest in her hips.

John quickly took and seized the moment and applied pressure. He said, "So, are you coming or not?" Natalie couldn't resist the temptation and then quickly decided to take him up on his offer for a cocktail at the watering hole. Natalie told him, "I will go but I am only going for about an hour because I need to be getting home because my husband has to be at work shortly."

Once they arrived at the tavern, John looked around for a secluded booth so that he could sit right close to Natalie. Once he found a booth, John did what he did best. He served up his slick sweet shit talking. Natalie was captivated by his cheap charm, and soon after taking a seat, John also began rubbing Natalie on her back.

John anxiously leaned over, giving a succulent, moist, wet kiss to Natalie's neck. Natalie was immediately swept into his lustful abyss after a couple of drinks and was captivated by John caressing her. Natalie closed her eyes while enjoying the attentive attention she was receiving, but once she opened her eyes, she glanced over and got a view of the wall clock. Natalie realized that she was already late getting home in time for Bash to make it to work on time. She jumped up from the booth and said, "Dammit, I've got to go!"

Natalie told John, "Since we are off tomorrow, we should get together." John smiled with lust still on his lips. He rubbed the palms of his hands from the front of his forehead to the back of his head. John said, "Yeah, darling, that sounds good." however, John was disappointed because he wanted her that night. John realized that it was easier than he had previously predicted. John said, "Come by my place tomorrow, and I will make us some breakfast while dishing up a devilish laugh." John gave Natalie his address and then one last wet kiss to her lips before she left the tavern. John wanted Natalie's mind to stay on him for the rest of the night. He once more leaned back into the seat to watch that switch walk out of the tavern. John lustfully said to himself, "Umm… umm… umm…," then looked at his watch to count the hours before he knew that he would ravish her.

Chapter 11: Sneaky Links

Once Natalie arrived back home, she saw that Bash was very upset because she had made him late for work. Bash said, "Natalie. What took you so long? You know what time my shift starts?" Natalie had already put a lie together on her way home. While stuttering, she said, "We, uh, had to finish prepping operating rooms that need to be ready for surgeries first thing in the morning, and I left without even finishing. I will be going in to work tomorrow early in the morning to get them finished even though it is my scheduled day off. I'll take Andre over to my mother's house for her to watch him while you sleep."

Bash believed he smelled alcohol on his wife's breath as she was explaining why she was late. Bash surprisingly asked, "Have you been drinking?" Natalie quickly jumped to her defense, and she responded in a disbelieving tone. She said, "What?" to make Bash think that he was imagining smelling anything on her breath just to throw him off.

Natalie finally said, "Listen, Bash, don't start that shit with me." to end his questioning. "I just got off work, and I am too tired for any of this silly nonsense." She then went into the kitchen to grab herself some of the leftover food that her husband had prepared for dinner so that the smell of alcohol would not be as strong on her breath.

The next morning, Bash arrived back home at 8 am. He found that Natalie was already up with Andre and had already fed him, dressed him, and was ready to go over to his grandmother. Natalie made sure that her hair and makeup were perfectly done. She spent so much time running back and forth from their bedroom to the bathroom that it eventually caught Bash's attention.

Bash became curious as to why his wife was working on so much detail about her appearance if she was only going to work. That was when he asked her, "Why are you going through all of that just for work?" Once again, Natalie immediately jumped to her defense. She said, "I don't know what you're talking about because I do this every day." Natalie said in a nasty tone, "Look, Bash, I'm a grown-up. Just know that I can do whatever in the hell I please, so stop questioning me!"

Bash responded and said, "In the first place, Natalie, I am not trying to dictate to you, but, surprisingly, I haven't ever seen you go through this much detail getting ready to go to work before." Natalie dared not turn around to look at Bash. She knew that if she had done that, Bash would see all the lies on her face.

Once Natalie went to drop Andre off at her mother's, Mrs. Montgomery also noticed how extra nice her daughter looked. Mrs. Montgomery asked, "Natalie, so where did you say you are headed again?" Natalie said, "Now, here you go, sounding just like Bash! I can't believe you both are acting as if I look like some 'cat' that's been drug through some 'greasy' alley!" Mrs. Montgomery looked at Natalie, knowing that she had seen and heard that behavior from her daughter before. Mrs. Montgomery said, "I just hope that this look is for your husband, but just know that cats don't only get drugged through greasy alleys. They do also get drugged through slick crap!"

Natalie rolled her eyes up in her head, but she made sure she kept it out of her mother's view. She then rushed, leaving to be on her way to John's house. Natalie drove fast, hoping to arrive at his house and hoping not to be seen by anyone. Natalie knocked at John's front door in a hurried manner and for him to rush to let her in.

John made his way, then opened the front door, and with a slurred-sounding speech, he said, "Ok, damn. Slow down, girl. Shit, who are you running from?" John, by that time of the morning, already had a cocktail in his hand. His shirt was halfway unbuttoned, showing the beads of hair on his fat chest. John licked his lips, then took another sip of his drink and said, "Come into my castle."

Natalie walked into the living room and then looked around at the shabby, unmatched furniture. She heard slow, sensual music playing and smelled incense burning. John hurried back into the kitchen, and that's when she noticed him walking on the backs of some dress shoes as house slippers. John went to finish frying up some sausage and potatoes. Natalie stood in the kitchen doorway and saw what looked like every dish that he owned piled up dirty in the sink.

John said, "Come on over here and have a seat while I finish cooking. I want to make sure that you eat and get everything that you came for," with a sly ass grin on his face. Once John finished preparing breakfast, he prepared Natalie a plate and brought it over to her as he leaned down and said, "I need to finish that kiss from last night." John gave her one sloppy, wet, alcohol-filled kiss on her lips as soon as they had finished eating breakfast. John took Natalie by her hand and led her away from the kitchen towards his bedroom. Natalie followed him into his bedroom with excitement. She saw that his bed was dressed in black satin leopard décor, and the room looked as if he hadn't made the bed in days. The bedroom had alcohol bottles and glasses scattered all around it.

Natalie said, "Damn, John, you really need to clean and fix up." John said, "Well, now that I have a woman coming to my home, I expect that's going to be changing soon," as he laughed. John smacked Natalie on her butt as if to say, "Good job for looking out." John began kissing Natalie and undressing her. He told her that he couldn't any longer resist those wide hips that kept calling his name.

Natalie was blushing like some in-love virgin schoolgirl. John anxiously hurried to undress his chubby, beer-bellied body, then climbed into bed with Natalie. John looked over Natalie's curvaceous body as he began to caress her breast. He then made his way down to her stomach and began kissing her between her thighs, making her tremble as Natalie wrapped her legs around his neck.

John, with a braggest tone, told her, "Yeah, that's right. Let me get all this wide wonder!" Natalie liked him controlling her. She started moaning from him, giving her sensual erotic pleasure with his tongue. John stopped for a moment, then said, "I'm not going to be done until you are calling my name."

John took his husky hands and parted her thighs even wider. He climbed his large body between her legs, and without waiting, he roughly penetrated her, already awaiting wet warmth. It was with each rough, demanding, deep stroke that Natalie moaned, not from pleasure but from his control of making her take it how he wanted to give it. Natalie began calling out John's name over and over, letting him know that she was submissive to him for his needs. The entangled, sweating sexual session lasted for a couple of hours before they were back at it once again.

Natalie noticed that their time for the ending was soon approaching, but not before she said, "Baby, I need to be getting home." John looked at her with shocking wide eyes and a questioning scowl on his face while thinking to himself, did this bitch just call me baby?

John immediately realized at that very moment that she was already hooked on him. He asked, "When am I going to see you again?" Natalie responded, "I'm going to tell my husband that I'll be picking up some extra overtime hours so I can see you more often. I'm making it look like I'm working overtime every other day before my regular shift so that we can get together. What do you think?"

John said, "Yeah, whatever you need to do, but what I do know is that I sure did enjoy them hips." Natalie turned, looking back at John, and sheepishly said, "I bet you did, but right now, I need to get a quick shower and then straighten my hair before I pick up Andre from my parent's house. My mother and Bash both made it a point to mention my appearance this morning, so I need to return looking like I did when I left."

Chapter 12: The Slicker They Come

Ronnie, who was like Bash's brother, was on his way to work that same morning shortly after Natalie had arrived at John's place. Ronnie always traveled the same route daily, which was something that no one knew. As he was approaching John's premises, that was when he noticed Natalie's vehicle parked but he didn't think very much of it at first.

A couple of days later, Natalie arrived again at John's house for another session of 'good timing' between the sheets. Natalie made sure to keep her schedule the same so that it did not raise any suspicions with her husband or her mother because she had learned from them to not overcompensate for her looks from the two days before. Natalie made sure she arrived back home on time, so Bash would not be late for work.

One day, Bash was telling Natalie how he appreciated her efforts and how proud he was about her helping to make their dreams come true faster by working overtime. Natalie just nodded her head, acknowledging his comment but with the corners of her mouth curled up.

Two days later, Ronnie left a few minutes earlier for work and took his daily route. Once more, he noticed Natalie's car, but this time, as he looked towards her car, Ronnie saw Natalie standing in a doorway, embraced by a man whom he did not recognize. Now, what he did see was both of them tongue-tied to each other, seriously swapping saliva.

Ronnie could not believe his eyes. He immediately thought of his brother Bash. Ronnie was infuriated after seeing them; he first thought about honking his horn to get Natalie's attention, but he then thought, I should roll this window down and scream, "**Hey cheap one, what in the fuck are you doing**?" Ronnie's mind went racing with angry thoughts after seeing that Natalie was having an affair.

Once Ronnie eventually arrived at the boxing gym, he sat at his desk contemplating picking up the phone and calling Bash, but he was unsure of what he would even say to him. Ronnie sat for a while and thought of several plans, but all he knew was that he did not want to verbally deliver that kind of painful news to Bash.

After a couple of more days, Ronnie was making his way to work, thinking to himself, 'This cheating ass **broad** better not be here!' Ronnie slowed, looking to see if he would see Natalie or the car, and just like clockwork, there sat her vehicle.

Ronnie realized that he had to do something fast because he loved Bash and was a loyal friend to him. Ronnie thought about calling and talking to Natalie, telling her exactly what he knew, but he also knew that it would end up in one ugly altercation between them all and with her lying to Bash, causing problems between him and his brother.

Ronnie was more fearful of losing his friendship with Bash because they had known one another for a while. He made a final decision and then sat down to type a letter to Bash explaining some details about the affair and sent it to his sister's home. Ronnie felt it would alleviate the risk of him losing his friend or Natalie intercepting the letter if he sent it to their home. Ronnie typed the letter without noting who the sender was, then mailed the letter to Gia's house, making sure to address it solely to Bash so that Gia would not also open it.

The following day, once Gia heard the mail dropping through her mail slot onto the floor, she sorted through the mail, and that was when she noticed a letter addressed to Bash. Gia removed her brother's letter and laid it on the kitchen table for whenever he made his way by for a visit.

When the weekend arrived, Natalie told Bash that she was going to work for a few hours and that she would be back after she was done. Bash then got up after Natalie had left. He dressed Andre and then went to visit his sister. Gia was glad to see Andre and her brother. After about an hour Andre began to whine for a bottle, so Gia said, "Let me heat a bottle because he's probably hungry or ready for a nap."

Gia went into the kitchen and remembered that Bash had a piece of mail there. Gia called out to Bash. She said, "You have some mail in here on the kitchen table. It came for you about a day ago." Bash said, "That's strange because I have never used this address before for anything."

Bash immediately got up and placed Andre into his walker. He then went into the kitchen to see what Gia was talking about. Bash looked at the letter front to back, then slid his finger under the sealing label, opening it. He began to read what it was telling him as his heart started racing, and he felt like his adrenaline went full pump.

Gia turned around with Andre's bottle in her hand. She was about to ask her brother to check if it was warm enough, but when she looked at her brother, she saw that it was as if he had received some sort of devastating news. Bash's mouth was wide open, and his eyes were wide. Bash was frozen. He could not even move. Gia kept asking him, "What is it, Bash? What's wrong?" Bash could not even speak, so Gia took the letter from his hand to see what was wrong with her brother.

The letter read, '*Your wife is having an affair with this man who lives at 1231 Mara Largo Avenue. You will find her there every other day around 11 am. Your wife's car is always parked in front of his house. The man's description of your wife's lover is that he is a light-skinned, heavy-set man with sandy brown colored hair as I have seen them kissing mouth to mouth. Trust me when I say that it was no cousin kiss, and you need to check it out for yourself.*'

Gia asked her brother, "What are you going to do?" Bash appeared to be dumb-founded. He said, "I don't know, but I need to take Andre and go right now!" Gia said, "I am here for you, Bash, but please don't do anything foolish until you do some investigating. Bash, please remember what Dad and our big brother always said: "That it's not what you know, but it's what you can prove."

Chapter 13: My Brother's Keeper

Bash finally arrived home with Andre, and once he got him down for a nap, he gave Ronnie a call. Ronnie saw that it was Bash on his caller ID, and he became nervous because he knew he should have received that letter by then. So, he quickly answered and said, "Hey man, what you have been up to?" while trying to sound normal. Bash just said, "Man, I need to come by this evening to talk to you. Is that ok?"

Ronnie knew that Bash had received the letter because he heard the despair in Bash's voice. However, he went on to say, "Yeah, man, no problem, I'll be here. What time are you coming?" Bash said, "I believe that Natalie should be arriving home soon, that is if she does not stop off somewhere."

Ronnie knew exactly what Bash meant by his last comment. He was sitting in the living room, letting the television watch him while thinking of how to handle the situation. Natalie finally arrived home. She walked in, and when Bash saw his wife, he had a strong look of hurt on his face.

Natalie, with agitation in her voice, asked, "What are you sitting there, looking silly in the face for?" Bash knew it was best not to respond and say nothing because he knew it was better not to accuse anyone without factual, solid evidence.

Bash's brother Jacques Bouvier Jr. Esq, always told his siblings what his father and his law studies had taught him: 'It's not what you know, but rather it's what you can prove that matters!' That was what kept ringing in Bash's head. He got up from the couch and then told Natalie that he was going to go visit Ronnie.

Natalie asked Bash if their son had eaten. Bash said, "Of course, he has eaten, Natalie! I stay loyal to my family, you know." while staring at her. Natalie didn't know how to respond to Bash's comment. His comment caused Natalie to sense that something was not quite right, which gave her the chills.

Natalie was wondering if Bash had a clue about her and John, she then thought, there's no way he knows! Immediately, Natalie decided she needed to scheme up a new plan just in case her husband was onto something regarding the affair because it was just getting good to her.

Bash arrived at Ronnie's, and when he walked in, Ronnie was certain that Bash was going to see a look of exposure on his face as the one who had written the letter. Ronnie kept his conversation with Bash limited by only saying, "What's up, Bash?"

After seeing Bash's look of hurt on his face, Ronnie said, "You sounded strange on the phone when you called. What's going on?" Bash said, "Man, you are not going to believe this shit!" Ronnie said, "Come on, have a seat. It looks like you might need a stiff alcoholic beverage, but I've got beer if you want that, too."

Bash said, "Absolutely, let me get a beer." Ronnie returned with a strong alcoholic beverage for himself and a beer for Bash before they took a seat to talk. Bash began the conversation. He said, "Check this out, man.

I was over to Gia's this afternoon for a visit, and someone had mailed me a letter to Gia's home, and right now, I don't know what to think."

Ronnie asked, "What was it? Because it seems to have you shook." Bash said, "Yeah, it does, but hold on, I've got it right here in my jacket pocket. I'm going to let you read it." Bash pulled the letter from his jacket pocket and handed it to Ronnie.

Ronnie took the letter from Bash's hand and read it like he hadn't typed it and acted shocked. Ronnie had all the dramatics in his voice. He said, "Man, damn! What the hell do you think about this letter, or better yet, what are you going to do?" Bash responded, "I really hope that it's just some bullshit and whoever sent it is playing a horrible joke."

Ronnie again asked, "So, what are you going to do?" Once again, Bash did not respond. Ronnie said, "Listen, man, the damn instructions told you to go and that you'll see it for yourself." Ronnie said, "Sounds like it has a solidity to it to me, so I don't know what else to tell you."

Ronnie was already at his limit with it because he knew the truth. He said, "Look, Bash, first off, fuck all that guessing shit! Bash, you need to go like it said now. Are you going to stop past there or not?"

Bash finally said, "Hell yeah. I'm going because it's like I don't really have a choice now that they have aroused my curiosity!" Ronnie asked, "Can I ride past there with you?" Bash said, "OK, because I need to see this for myself."

Ronnie anxiously sprained up from his seat and snatched his jacket off the chair without missing a beat. Ronnie said, without hesitation, "Let's head out over there." He asked Bash, "If we see this fucker, oh shit, my bad Bash. Please believe me, there was no pun intended in that, but please just let me be the one to skin his motherfucking ass up. It will be my pleasure!"

Bash stood there in the center of the floor for a few minutes looking like he was in a daze after Ronnie had stated that he wanted to beat the man's ass. Bash then quickly remembered and said, "You tried to warn me about my wife, and so did Gia. I guess I was foolishly blind and wanted to see the good in Natalie." Ronnie looked side-eyed at Bash but did not speak out his thoughts about her at that moment. What Ronnie really wanted to say was, 'What damn good?' but he didn't want to disrespect Natalie to Bash.

Ronnie was standing at the front door with his jacket on and his hand on the doorknob before his patience had run thin. Ronnie raised his voice and then said, "Bash, man, bring your emotional ass on here instead of standing there staring at the ceiling. Besides, it's going to be getting dark soon, and I want to make sure I lay his ass down on the curb to go to sleep before we leave him."

Bash and Ronnie finally reached the destination of the address that was mentioned in the letter. They parked directly across the street from John's house. Bash had high hopes of seeing the man who was described in the letter. They ended up sitting there for a couple of hours but eventually figured that he might not be at home because no one ever showed a presence.

As soon as they were about to leave, John pulled up in one old rust bucket of a car. John parked directly in front of his house and got out of the car. Ronnie saw that it was the man that he had seen Natalie swapping saliva with.

John got out of his car, and it took everything within Bash not to get out of the car. He badly wanted to confront him and stomp a new mud hole in John's behind. Ronnie anxiously said, "That's gotta be him because he's going up to the house that the letter implied where he lives. Besides that, he sure does fit that shitty-ass description."

Bash shockingly said, "You mean to tell me she's really been having an affair with that piece of shit," he then went on to say, "Wait, I'm just trying to figure out what in the hell she sees in this scrounge raggedy-looking fat fucker." Bash couldn't take his eyes off John; he was paying close attention to everything about him. He even took close notice of John's hair color.

Bash, in a disbelieving tone, said, "Just look at this horribly unattractive bitch, looking like he's swollen to the damn colon. Wait, is that tumbleweed upon that fucker's head that he's trying to pass off as hair?" Bash's comments caused Ronnie to crack up laughing. Right then, Ronnie joins in on the pickapart of John's appearance.

Ronnie said, "Man, hold up! Bash, please tell me how in the hell did you miss those crooked-over-ass shiny Sloafers he's sporting!" Bash asked, "Ronnie, what in the hell are Sloafers?" while laughing. Ronnie could not wait to respond while laughing. He said, "Hell, I don't even know, but what I do know is that them sure ain't no fucking loafers that his bitch ass is sliding around here in." By that time, John had noticed the commotion that was going on in the car while they were looking at him laughing. John quickly retreated to his house but not thinking much about the men in the car, other than it just looked strange to him because he had never seen them before.

Bash began talking out loud. He said, "Now, what she has been telling me is that she's been going to work early every other day for overtime, but what she has really been doing is coming over here lying up with this piece of shit!" Bash went on to say, "It's making sense to me now. No wonder she has not been wanting me to touch her. I know it's been over a couple of months since we've been intimate. Meanwhile what she's been telling me is that she is too tired and that it's stemming from working all those hours and taking care of Andre."

Ronnie impatiently asked, "So now what?" Ronnie told Bash, "Let's just go on up to the house to confront this motherfucker because you know I'm down for whenever!" Ronnie started sliding on his leather gloves to protect his knuckles. Bash said, "Not right now, Ronnie. First, let me go home to confront Natalie's cheating ass; I need to see her reaction but trust me, I'm coming back for this bitch!"

Ronnie got agitated and said, "What?" in a frustrated tone. Ronnie became upset because he didn't get the opportunity to beat John's ass the way he had intended to. Ronnie took in a deep breath and then shouted out loud, "Shit, I can't get nothing!" That angry comment Ronnie made caused Bash to laugh.

Bash said, "Man, damn, if you don't calm your ass down, hell, you don't even have a dog in this fight!" The only thing that Ronnie had left in him was to sarcastically say, "Yeah, but my dog does." He gave a big smile to Bash. Bash chuckled and said, "That's my dog!"

Chapter 14: Richter Scale

Meanwhile, Natalie had already called and talked with Adrian, her sister. She told her all about John, with every bit of lustful excitement in her voice. Adrian said, "Natalie, you really need to be careful before someone gets hurt." Natalie said, "I know because when I arrived home today, Bash was acting strange, and I don't know why?" Natalie told Adrian, "I think that I should change things up from how I've been doing it because I'm not ready to stop fucking John just yet. I think what I'm going to tell Bash is that the job is cutting the overtime hours, which means that John and I will just have to get even more crafty in getting together, possibly only on the weekends. I'm also going to tell Bash that you and I will be going shopping on Saturdays, and then I'll take Andre over for Mama to watch for me. I think that might work. Adrian, what do you think?"

Adrian said, "Listen, please leave me out of your affair. Besides that, I don't want no parts of it or when your husband finds out, and you get caught! Natalie, just know that you are going to slip up trying to be sly and remember that I tried to tell you." Adrian continued, "Your husband loves you, and I just don't get why would you do this to Bash." Natalie frustratingly said, "Well, let me think about it, and I'll call you back later!"

As their phone call came to an end, Bash arrived back home, and his wife could see that he was more visibly upset. Natalie, in an upbeat manner, said, "Bash, why don't we go to the drive-in tonight?" Bash looked at Natalie with complete disgust, then in a sarcastic tone, he said, "I am definitely not in the mood, but I'm sure you understand how that is."

Natalie looked at her husband, then tried to pour a sweet venomous serving up to him by saying, "We don't get to see each other very much anymore, so please amuse me just this one time."

Bash looked eye to eye with his wife when he boldly said, "Here's an idea, why don't you get your fat ass sandy hair boyfriend to take you?" Bash had a stone-cold face after he made his comment to his wife. Natalie was utterly surprised, and without haste, she said, "I don't know what the hell you and Ronnie were over there drinking or doing, but that's some fucked up shit to even say to your wife! I also need you to know that if I had a boyfriend, you can best believe that I would not be here with your stupid behind!" Natalie then went in deep and said, "Bash, you can let whoever is running some damn lies down to you be the cause of you fucking up your marriage if you want! I also would advise you to listen because I am only going to say this one time. You better know that I will take my child and move the hell out of here; now, keep acting like a damn fool if you want!" Natalie finished her conversation with, "Now I'm taking Andre, and we are going over to Adrian's for a while, but you can call whatever bitch that done feed you that bullshit. However, just make sure you tell that lying ass bitch that they can kiss my ass in the process!"

Bash sat dumbfounded, in deep thought, after Natalie left the house because she sounded convincing to him. Bash had a couple of stiff drinks to calm himself down. His thoughts were racing and leading him into dark, confusing places. Bash kept trying to figure out who he thought would know how to send him anything at his sister's house. He came to one conclusion and thought absolutely no one. What dawned on him were the things that his wife had told him concerning his sister with not liking her.

Immediately, Bash thought the worst of Gia. He believed that she had concocted the entire story, then typed the letter and mailed it to her own home. Bash suddenly believed all the things that Natalie had been telling him regarding his sister and that it had to be the truth.

The first thing that Bash watched for the following week was to see if Natalie would be leaving early for work on those scheduled days that were mentioned in the letter. Once he saw she was not leaving on the scheduled days, as mentioned, he felt even worse about jumping to the conclusion and accusing Natalie of having an affair.

Gia phoned her brother after a few days to check on him, and that was when Bash angrily said, "Gia, I cannot believe that you could have stooped as low as you did! I realized that you were the one who created that trash-ass letter about my wife, but then you acted as if someone had mailed it to your home. You could have ended my marriage with that foolishness. Once I really examined that letter, I realized that it was addressed to Bash. I realized only my family calls me "Bash." That's where your scheme fell apart and unfolded in front of me. Gia, I also understood that man could have been just about anyone that you may have noticed just by riding down the street, but then you put an innocent man's life in danger. I am even more upset that you chose such a slouchy-ass-looking man. I should have understood that once I saw him."

Gia could not believe what she was hearing, and that was when she said, "You must be out of your damn mind to think that I typed that letter and then mailed it to my own home. It is completely insane, but I also understand who you are sitting around listening to as well!" Bash said, "I'm struggling with this and I really don't want to talk to you any longer right now! Gia. Ronnie and I had gone and sat out in front of that man's house, and once I got a view of him, I wanted to kill him."

Bash, with an irate tone, said, "I need to get off this phone. You need to give me time to be able to process and fix this shit that you've caused! Listen, please don't call me. I'll call you when I am ready to talk. Gia, just know that I never would have expected this out of you."

Natalie went for a visit to John's house the following day after Bash had accused her. She also took Andre with her. Once Natalie arrived at John's and entered the house, she excitedly said, "We need to talk. I came to let you know that we are going to have to be more careful because my husband accused me of having an affair yesterday evening, and what's crazy was how he described you."

John, with a surprised tone, said, "What?" He got nervous as he began stuttering over his words. That's when he asked, "How would he know anything about me or what I look like." John quickly remembered the strange men in the car he had seen. He said, "You know what? I pulled up to the house after doing some running late yesterday evening, and there were two guys sitting in a car across the street from my house watching me. I knew it felt strange when I looked over at them because they were acting hostile one minute and then laughing the next! I did not recognize either of the guys because the windows were slightly tinted, but something told me to take my ass in the house! Now, once I got into the house, I was peeking out of the front door blinds to see what they were doing, but I still did not get a good look because they left shortly after a few minutes. Damn! I did not know my life was that close to danger."

"But you handled it right?" Natalie said.

"I did."

"Besides, I'm not worried about Bash. I can handle him, but it's that damn Ronnie's ass that we need to be more concerned with!"

Natalie asked, "What kind of car was it?" John described the car in full detail, stating that it was a black four-door 750 BMW with tinted windows. That's when Natalie realized that it was Ronnie's car. She knew that Ronnie and Bash were together that evening because of how her husband had described John.

Natalie nervously said, "Oh shit! Let me get out of here! I will see you at work because it is not safe for me to come back over here." John nervously said, "No, it's not! I don't need any troubles in my life, and besides that, I would hate to leave your child fatherless!" That comment did not even detour her decision to still want to be with John.

Natalie hurried and got back to her car without even buckling Andre into his car seat, then jumped back into traffic, rushing to Adrian's house. Once she got into the house, she laid Andre down as he slept and said, "Adrian, girl, he knows," sounding like she was out of breath.

Adrian asked, "Who knows what, Natalie?"

"Bash knows about John and me. I just left John's house, and he said that there was a car parked in front of his house yesterday evening and also that two guys were watching him."

Adrian asked, "OK, but how do you know that it was Bash?"

"I asked John what the car looked like, and he described it in detail. It was Ronnie's car! I know when Bash left the house yesterday after acting all strange, he said that he was going to visit Ronnie. Now you know that fool Ronnie is with all the drama when it comes to Bash. Besides, he said there were two guys in the car who he did not recognize. Adrian, it was them because Bash came home then accused me of messing around, and he described John. Adrian, I knew that something was off with Bash yesterday. Now, what if I had gone back over there?"

Adrian said, "But you did go back over there even after you thought that Bash knew yesterday. Now, what I can't believe is that you took Andre with you, which was way out of line for you to do, but remember, I told you that it was just a matter of time."

Natalie said, "I know, but I only went to tell John that I could not come back over there anymore. I'm going to see him at work in a couple of days, so I will let him know what the new plan is from now on."

Adrian could not hold her judgment in any longer. She then said, "Damn, Natalie, does he have you that strung out?"

Natalie sighed, rolling her eyes up in her head and closing them, but not before she did her usual aggravated gestures of sucking her teeth. Natalie didn't hesitate to remind Adrian to not forget that on Saturday, she was going to be telling Bash that they would be going shopping together. She knew this would work so that she could meet up with John. Natalie also told Adrian, "Please remember that I am going to tell Mama the same thing, just in case she decides to ask you." Adrian angrily said, "Look, you need to stop digging that deep shitty ass crater that you're already standing in and leave me completely out of it."

Ronnie was once more on his way to work. He expected to see Natalie's car parked in front of John's house. He slowed down, looking for either of them, but to his surprise, she was not there. Ronnie figured that Bash must have said something to Natalie like he said he was going to do.

Each day thereafter, Ronnie looked to see if he was going to see her car parked there, but a couple of times when he drove by John's house, he happened to see John out on his porch. John saw him each time and remembered that it was the car that was parked out on the street that was watching him.

One time, when Ronnie rode by and saw John, it caused him to remember how John and Natalie were tongue-locked together. Ronnie slowed down, staring John directly in his face, wishing and hoping for any sort of aggressiveness of a spoken word directed towards him. Meanwhile, he was praying to himself, "Please let this bitch make just one aggressive gesture so that I can get out of this car to beat the shit outta him!" John saw the terror in Ronnie's eyes and remembered what Natalie had stated regarding Ronnie that he was crazy. John knew immediately that he did not want any parts of him. John waited until Ronnie had passed by him before he said out loud, "This motherfucker is crazy!"

The weekend arrived. Ronnie called Bash to see what he was doing for the day. They decided to go on a golf outing together. As they were riding on the cart, Ronnie asked Bash, "So, how are things going between you and Natalie?" Bash said, "Ronnie, man, let me first start off by apologizing for dragging you into what could have turned out to have been a horrific mess."

Ronnie said, "Man, no need to apologize, Bash. You are my brother, right?" Bash said, "Absolutely, you are my brother's keeper, but listen, when I arrived back home, I lit into Natalie's behind after she asked me to take her to the drive-in because we hadn't spent much time together in a while. I immediately told her that she could get that sorry-ass-looking sandy-headed fat fuck of a boyfriend to take her. Ronnie, she damn near went into hysterics. She said if she had a boyfriend, she would not be with my stupid ass, Ronnie. She also said that she would take Andre and leave if I kept on with the accusations. I just remained still for the rest of the evening, trying to think and to calm down. I'm honestly glad that I did because I wasn't thinking clearly before I had gone back over there and killed that man. Natalie was so mad that after she came back from her sister's house, she took Andre and stayed in his room with him for the night. I guess so she could calm down, I sat there, then had a stiff drink or two and kept examining that letter, but that's when I ultimately realized that no one would know how to send me anything at my sister's house other than family. Natalie had been trying to tell me that Gia had a dislike for her a while ago, Ronnie. I realized that Gia was the one who had typed that letter and then mailed it to her own house and waited for me to come over so that she could give it to me as if it were from some anonymous sender. I am so damn mad with her right now that I don't know what to do!"

Ronnie said, "Come on, Bash, now you can't believe that your sister would do that!"

"Ronnie, look, all I know is that Natalie has been at home with us every day until she has been scheduled to go in for work, regardless of what that letter stated but to answer your question, yes, I do believe that Gia did that shit."

Ronnie said, "No, Bash, don't be mad with your sister. She did not do it; blame me because I was the one who sent it."

Bash said, "Good for trying to take the blame, but I don't want to even talk about this any longer. Please!"

Ronnie said, "No, wait, I'm telling you the truth. Just let me explain." But Bash refused to listen and said, "Please stop before I get upset with you too."

Meanwhile, Natalie had gone to drop Andre off for her mother to watch, and once inside, she told her mother that she and Adrian were going shopping. Natalie didn't know that her mother already knew that Adrian was at work because they had just talked to each other. Mrs. Montgomery said, "This is the last time that I'll be watching Andre, and you need to find you another sitter while you go shopping." Mrs. Montgomery was looking directly at Natalie's face so that Natalie could see that she knew she was lying to her.

Mrs. Montgomery said, "I assume you're going shopping looking for a lesson when you already have a blessing, huh." Natalie ignored her mother's comment because her mind was on getting to John. She did not want to debate, knowing that her mother might tell her to take Andre with her right then.

Chapter 15: It's About to Hit the Fan

Natalie and John met up at an out-of-the-way motel for a few hours. He said, "Natalie, listen, after this, we need to be cool because I don't need your husband looking for me." John was not able to stick to what he claimed would be the last time, partially due to Natalie wanting to get together with him. Natalie had even started taking her son with her instead of getting a sitter and then putting him to sleep while she and John carried on with having sex.

The affair carried on for a few more months, but it suddenly ended due to John seeing another attractive woman that the hospital hired for the housekeeping department. John instantly grew tired of Natalie and decided that he was done with all the sexual escapades with her. Without a warning, he was onto his next sexual partner.

Natalie had gone to the watering hole after work one evening to see what John was up to because she noticed that he was no longer giving her any attention. That was when she saw John cornered off with the new hire and pulling the same slick antics that he had pulled to get her. Natalie was irate when she saw him. She walked over to the booth, picked up his glass of cognac, and threw it in his face. Natalie angrily said, "You are nothing but a lowdown ass dog, you no good son of a bitch." She then turned and strutted out of the tavern, mad as hell thinking that her comments mattered to John.

John said to the new hire, "That woman has a mental problem." John sat there with a wet-lipped side-mouth grin, then wiped the drink from his face as he beckoned for the waitress to bring him another drink. John continued with his conversation as if nothing had ever happened.

The next day, when Natalie arrived at work, she could not wait to give John a piece of her mind. However, she was in for another surprise that she was not going to be able to handle. Natalie looked at the schedule and found out that she was no longer partnered up with John. Now what the new schedule read was that John was now partnered up with the new hire.

Natalie could tell that she had been the talk of the housekeeping department because it seemed that everyone steered far away from her. She also noticed that no one would even make eye contact with her. Natalie understood that it was stemming from how she had behaved at the watering hole the night before.

John made it a point not to even take any of his breaks at the same time as Natalie, just to avoid any sort of confrontation with her. Natalie was more than hot that she was not able to give him a piece of her mind and that she could no longer enjoy sex with him.

John would not answer his phone when it rang once he saw that it was Natalie calling him. She eventually stopped trying to reach him after about two weeks and once she arrived at work, she then shamefully kept to herself.

Chapter 16: What's Done in The Dark

The following month, Natalie missed her menstrual cycle; she realized that she had become pregnant, but this time, when she told Bash that she was pregnant, he instantly had factual evidence regarding his wife's cheating. Bash remembered that he and his wife had not been sexually active in several months. Therefore, he knew there was no possible way that the child could be his. Bash realized at that moment that Natalie had been having an affair. He then remembered how he had blamed his sister for lying regarding his wife and how he scolded Gia as harshly as he had.

Bash was livid. He said, "Natalie, please tell me how in the FUCK can you be pregnant if we have not been sexually active, that is, at least together, for some time now?" Natalie came bold, already prepared with another lie. She said, "Bash, just because we have not had sex for a few months does not mean that I was not already pregnant!" Natalie was grasping at straws, trying hard to convince her husband. She then told him that she had a short period the month when they were last intimate. Bash usually believed just about whatever she told him for at least until then.

Bash said, "Natalie, well, there is one thing that is for certain and that is I know for a fact that I am no fool!" Bash went as far as to say, "I just look like this. Besides that, your shit is not adding up, nor is it making any sense." Natalie's only shady defense was to say, "You must be crazy. You will see once this baby gets here." Bash said, "Now, for once, you are speaking some truth, and you are correct, which I cannot wait to see."

The embarrassment was too much for Natalie to bear when she would go to work and eventually risking everyone finding out that she was pregnant and knowing that people would start talking. Natalie also wanted to hide the pregnancy from John. She knew that he would be able to put the pieces of the puzzle in place and figure out that it was not her husband's baby. Natalie thought that he might even try to stake claims on the baby because he had no children. So, she gave the hospital a two-week notice, and she quit the job before anyone found out that she was pregnant.

The seventh month quickly arrived, and Natalie had gone into labor. Meanwhile, Bash was at work when he received the call. He was trying to get to the hospital in time, but his wife had already delivered the baby before he reached the hospital. The Bouvier's second baby was born, and this time, it was a girl whom they named Nadia. She was light-skinned like her mother, with sandy brown hair, which neither Bash nor Natalie or any of the other family members had that color hair.

Natalie knew that Bash would know at first glance that this baby was not his. Bash walked into the delivery room after the birth of the baby. He walked over, looked at the baby from head to toe, and said nothing at first. Bash then looked towards Natalie but not directly at her. He said, "Natalie, your daughter is beautiful." Natalie was shamefaced as she began crying and tried calling out for Bash, but he never broke his stride as he walked out of the room without saying anything else to his wife.

Bash, unbelievably, somehow, found the strength to forgive Natalie. He then decided that he would hide her shame. Natalie believed that it was out of the love that he felt for her, but truthfully, it was out of fear of losing his family again. Bash did not want his family to know that he had not taken heed to Gia's warnings regarding Natalie. Bash and Natalie even named the baby after Bash's mother, giving her his mother's middle name.

This was now the Bouvier's picture-perfect family, and they had just bought them a new family home, or at least this was how they viewed it as being perfect.

Chapter 17: Indiscretion Reminder

Eleven months later, after giving birth to Nadia, the Bouvier's found out that Natalie was pregnant again because she had forgotten a couple of times to take her pills on schedule, but she doubled up the following days, which did not prevent her from becoming pregnant.

Natalie did not want to be pregnant again, and she and Bash had previously discussed that they did not want another child. The endurance of being unexpectant and unwantedness was where their "scapegoats" journey would begin. Let's just say the Bouviers were none of the happiest regarding the pregnancy and were determined not to have the baby.

Natalie discussed with Bash about them aborting the pregnancy, but it was too late for that option. Natalie then decided to take measures into her own hands. She even attempted falling down some stairs one day, hoping that it would end the pregnancy, but she failed in her attempt.

Natalie and Bash tried every measure that they could think of, even down to some old wise tales. However, the pregnancy just would not go away. The Bouviers had come to the reality that for each daunting task they had tried, nothing changed the outcome, and they failed to make the pregnancy go away.

The Bouviers had to deal with the fact that they were going to be living the remainder of their lives with another addition to the family. Natalie's ninth month arrived, and yet again, she had begun labor. She was angry about having to experience the pains of labor once more, and unfortunately, that would be this baby's first mistake.

After several hours of long, excruciating labor, Natalie finally gave birth to a healthy 8-pound baby girl who looked like those Bouviers in the early morning hours. The delivery doctor laid the baby upon Natalie right after the delivery, and as soon as she saw her baby, she began harshly complaining to the nurses about the labor pains. Natalie was repeatedly saying it was unbearable and screaming about how that baby had almost killed her.

The delivery nurses were listening to Natalie while trying to calm her down. There was, not once after the delivery, an instance where she seemed concerned or inquired if the baby was doing alright. The nurses and the doctor did not know what to think of the behavior, as they all looked baffled. One nurse looked at another nurse and questionably mumbled, "What the hell?" as the other nurse responded by shrugging her shoulders and shaking her head as if to say, 'I can't wait to be done with her ass.' The doctor had grown impatient with the complaining, and as the afterbirth was removed, she snatched off her gloves indignantly and said, "Congratulations," as she exited the delivery room.

The nurses understood that this manner of angriness was not like most of the mothers who they help during delivery because most mothers were usually eager and asked to see their baby or if the baby was alright. The Bouvier's baby girl came into the world, a beautiful shade of caramel, just like her father, and looked exactly like Bash's mother with a full head of black curly locks.

Every nurse in the nursery was drawn to see the baby; they all marveled at her beauty as the nurses took turns holding her. The nurses kept telling Natalie how beautiful they thought her baby was. She simply replied to the nurses, "That baby almost killed me!" The nurses whispered to each other, "Oh, this is the woman that the other nurses were discussing to not mention anything to her about the baby."

The only thing that Natalie did care to discuss with the nurses was how her two other children's deliveries had not been anything like that, stating that they were not that hard on her. Natalie went on to say, "My husband and I had hoped that this baby would be a boy, but oh well, here we go with this hollering puss," as she snickered. The nurses could not believe what they were hearing. They looked at each other with disbelief and hurried to get the baby into the nursery crib.

The Bouviers had not even taken the time to figure out what they were going to name their baby throughout the length of the pregnancy. However, they had to rush and name their baby right before it was time to take her home. They named her Christiana, who was the chosen scapegoat.

Chapter 18: Let The Scapegoating Begin

The Bouvier's two close neighbors, Tammy and Rena, knew that Natalie was about to be released from the hospital along with the baby. They knew from what Bash had informed them of earlier that morning. The ladies were eagerly waiting to greet and meet Natalie and the new bundle of joy.

The two ladies saw the car when it pulled up into the Bouvier's driveway, and they hurried over to have a quick peek at the baby. All the ladies could do, once they saw the baby, was to makeover Christiana. They told both Natalie and Bash how beautiful they thought she was. Bash told the ladies that they were welcome to come in for a visit, which they were excited to do, and then took his offer to visit.

The baby was dressed in a soft pink gown that had flowers around the collar. She was also swaddled in a soft pink baby blanket with flowers on it that matched her gown. Tammy recognized the baby's clothing that she had gifted to Natalie.

The baby was curled up on the couch beside her mother, so still with that fresh smell. After a while, Christiana began to squirm, making those whimpering, grunting sounds. The ladies figured the baby was probably ready for another feeding. They felt they needed to give them room for bonding.

That's when Natalie spoke aloud out of frustration and said, "Here goes this hollering puss!" Tammy and Rena looked at each other and became puzzled by what Natalie had just said about her baby. The ladies noticed that Natalie did not display attachment towards her baby, as most mothers do right after giving birth. They assumed that she was probably tired from giving birth and that she was not thinking clearly.

Tammy and Rena spent most of their visit listening to the story of how Christiana had almost killed her. Tammy and Rena did not want to wear out their welcome, so they decided to end the visit and return to their homes.

The ladies told Natalie that they would visit her in a couple of days and told her to get some rest. After a couple of days, the ladies went back to see how Natalie was managing the children as they had promised.

This time, when the ladies returned for a visit, they took the Bouvier's some prepared dishes so that Natalie would not have to worry about cooking anything other than settling in with her newborn. Rena was even more excited because she, too, had just given birth to her daughter, Nikki, a couple of months earlier than Natalie. Rena thought it was great that the babies would grow up as playmates, seeing that they lived next door to each other.

As the months passed, Natalie and Bash seemingly settled in with their new addition to their family, but in all actuality, that was the furthest thing from the truth. All they did was complain about how the baby cried too much and how Natalie was sick and tired of changing diapers.

One day, Natalie was outside with the children when a swift spring pop-up storm came pouring down. Andre and Nadia made it into the house with their father, but Natalie and Christiana weren't as lucky; they got caught in the storm. Natalie and her baby were about six houses down the street from their house. The stinging raindrops were smacking Christiana in her tiny face, which made her struggle to breathe.

The heavy rain straightened and smoothed out Christiana's curly locs all around her face. She was dripping wet as she began to scream and cry with fear due to the stings and not knowing what was happening to her, even though she was being held in her mother's arms.

Christiana tried to draw close to her mother's bosom like most infants do, but that quickly irritated Natalie. She was frustrated with the baby, all because her baby was afraid. Natalie looked Christiana in her face as she held her on her hip. She ground her teeth together with anger, then looked her baby in her face, shouting and telling her to shut up!

Natalie gave her baby a shaking jolt, letting her know that she meant exactly what she was saying. Christiana's head snapped back from the jolt, and her eyes widened. Christiana looked at her mother for comfort, but the sting of the rain was once more taking her breath. She closed her small, dark eyes from all that she was experiencing at that moment. Christiana was trembling from experiencing something that no infant could possibly understand.

Christiana was also frightened by her mother's treatment as her tiny bottom lip began quivering, and she cried out even harder. By the time Natalie made it back to the house with her baby, they were soaked to the bone. Natalie was mentally exhausted with Christiana, and once they entered the house, Natalie jammed the baby into her walker that was set in the living room.

Natalie walked into the kitchen to get a dry hand towel. She returned to the living room, where Christiana was sitting and crying in her walker. Natalie roughly swiped her baby's face and head with one quick swipe, startling Christiana with the rough swipe to her face, causing her to gasp for breath again as she continued crying loudly.

Natalie looked at her baby with disgust and then glared at her while Christiana cried, reaching for her mother. Natalie looked at her and once again yelled at her to shut up.

Natalie said to her, "You make me sick," as she walked away, rolling her eyes with clutched teeth. Andre and Nadia stood alongside the living room couch, watching their baby sister as she cried, her tiny arms outstretched, looking for comfort from anyone, but no one came. Andre and Nadia were looking as if Christiana was something strange. Natalie then called for Andre and Nadia to come to the dining room table to eat their lunch. As Natalie prepared their lunch plates, the two siblings sat sideways in their seats, watching and listening to Christiana crying in her walker, but again, her screams went unwelcomed.

Natalie periodically shouted out to Christiana, telling her to shut up. The battle went back and forth until the baby's voice weakened as she managed to cry herself asleep right there in her walker.

Christiana was exhausted because her head was left hanging forward as she slept. Bash finally appeared and came into the kitchen. He told Natalie that he was on his way to pick up his new golfing equipment. Bash got ready to leave. He then bent down, giving Andre and Nadia quick kisses on their foreheads. He lightly patted his baby girl on the top of her head as he walked out the door.

Once Natalie was done feeding Andre and Nadia, she went and snatched Christiana out of her walker, startling the baby and causing her to cry. The baby tried to snuggle into her mother's bosom once she picked her up. However, that was also met with further fury of rejection from her mother.

Natalie hollered again, telling her to shut up, as she roughly handled Christiana carelessly as she placed the baby into her crib. Natalie began snatching the cold, wet clothing off the baby. Christiana's tears streamed down the sides of her cheeks and back into her ears as she lay helpless on her back.

Now, between those long periods of silence from crying, she forgot to breathe. Christiana grabbed at her mother's clothing for comfort. Natalie aggressively snatched her baby's shirt over her head while Andre and Nadia stood by the crib, watching, saying, "Mommy, she is bad!" Natalie swiftly replied, "I know she is getting on my damn nerves!"

Tammy came to visit more often than Rena because her children were almost grown, and at least one of them had already left for college. Tammy noticed that Natalie was still not displaying any excitement or attachment towards Christiana, and her attitude remained as it did from the first day when she arrived home with her baby.

Tammy also paid attention to how Natalie always seemed to make such an excited fuss over Andre and Nadia, but when it came to Christiana, it was always met with negativity. Tammy wondered how Bash could sit and never open his mouth to his wife regarding her behavior towards their baby.

Tammy had grown angry with how Bash had done nothing and let this behavior go on without any concerns. Tammy said to Natalie one day after a negative response during a visit. She said it as if she was making a joke. She was serious. Tammy said, "You should let me have Christiana. That is my baby." Tammy went further to say, "You know, all we have are boys, and I could use more estrogen in the house." Natalie realized that Tammy was serious, but she did not let Tammy know that she knew because it would be an admittance of guilt.

Natalie responded, then said, "You would give her hollering ass right back to us!"

After that visit, Tammy often went to get Christiana as much as she could while pretending to give Natalie a break. Actually, it was for Christiana to receive a break from her family. Tammy's husband, Reed, loved Christiana; he and Tammy would spend their money on toys and clothes to keep at their home when Natalie and Bash would let them keep Christiana overnights, which became more often than they ever expected.

Christiana seemed to love being with Tammy and Reed. The baby loved it when Tammy would place her in their kitchen sink to bathe her. Tammy made those times very special for both of them by placing toys in the water while pulling herself up a stool so she could sit and talk to Christiana.

Christiana would smile and laugh at the splashing of the water, and Tammy loved to see how happy Christiana was when she was in their presence. Christiana loved Tammy and Reed's tender voices, they could see that it soothed her when she was around them.

Reed was always happy to walk in and see how happy his wife was just to be spending time with Christiana. Reed would always come and ask the baby with an excited tone saying, "What are you doing, little girl?" Just to see her light up.

Reed also knew that Christiana was going to give him that big grin that he loved to see each time when he talked to her. Reed consistently saw how happy Christiana seemed to be when she was in his and Tammy's company. Reed always told his wife that they needed to give her to us. He said, "I know there may not be any blood ties, but she is our baby!"

Reed came in from an outing one afternoon to find Tammy napping with Christiana when he looked at how peaceful his wife and the baby seemed to be together. Reed got caught up in his emotions and couldn't help himself thinking about Bash and Natalie's behavior towards Christiana. Reed wanted to say it out loud, but he softly said, "I don't know what in the fuck is wrong with either of those two silly motherfuckers," and shook his head out of frustration.

Tammy wasn't asleep. She was just resting and had heard her husband's outward frustrating comment. The comment drew tears to

Tammy's eyes. She looked at Reed, then said to him, "She is our baby, and she already knows it whether they like it or not!"

Natalie and Bash would dare not let Tammy and Reed have Christiana because it would reflect badly on their image. Their image was what mattered the most to them, mainly when it came to their house, their cars, and how they kept the children dressed, especially Andre.

Natalie never let Andre get dirty like the other little boys that he played with, and whatever Nadia wore for the day, Christiana had to be dressed exactly like her sister, never allowing her to have her own identity. The Bouviers were seemingly training Christiana to walk in the shadows of others and never wanted her to be set apart as herself.

The months were passing fast, and Christiana was growing quickly. She had not begun walking quite yet, but she was very attentive to the things that were going on around her. The Bouviers always set out on their porch, as so did many of the other neighbors on those warm sunny days, just to watch the children play. There was never a need for any babysitters because the neighbors looked out for one another.

Andre and Nadia were playing in the yard one afternoon, and the neighbors could not help but notice Christiana sitting in her walker. The neighbors could see the excitement on her tiny face as she watched her siblings running and playing; her tiny arms were moving so fast, hoping to catch them as they ran past her.

Christiana's grin showed her two teeth that had just begun breaking through to the surface of her gums, and her soft, curly locs hung around her face. The neighbors couldn't help themselves from stopping as they walked down the street or passed by in their cars. It was because they hadn't seen the baby since her initial arrival.

Christiana was born late in the fall, and no other neighbors had seen her other than Tammy, Reed, and Rena before the weather changed. Everyone congratulated Natalie and Bash as they stopped by. Each one said to the Bouviers, "She is such a beautiful baby!" One neighbor said, "She's looking much like you, Bash," and how blessed they were to have her.

Natalie and Bash struggled to respond after most of the comments; they just managed to change the subject over to Andre and Nadia, making sure to praise them and change the subject. Andre was a short little boy who seemed shorter than most of the kids in his age group. Andre hated the fact that the other boys never let him forget that he was the shortest of the group when they played together, and Natalie hated it.

Chapter 19: Identifying Moments

Natalie believed that Andre and Nadia could not do any wrong because they both reminded her of herself with how they acted. Andre looked like both of his parents. However, Nadia was a light-skinned version of her mother, sort of an average-looking kid with coarse, sandy brown hair.

Nadia always wanted and needed to be the center of attention, and she could not keep friends because she thought everything should be about her. Nadia had this bitter attitude and was always mocking people, which steered her friends away from her.

Time was going by quickly. The Bouvier's two oldest children were now in elementary school, and that was when Natalie decided that she was going to find a new job. Natalie went to ask Rena if she would babysit Christiana once she found a new job. Rena gladly accepted, watching Natalie's little girl so that Christiana and Nikki could play together.

Natalie landed a great job that matched her husband's income, which added to their already escalated images. However, it required Natalie to work a second shift. Natalie took the shift despite not being able to spend time with her children except on the weekends. Or she only saw the two oldest children while they were getting up to get ready for school. This left Christiana with her mother for a few hours before she went to take her over to Rena's to be watched.

When Sundays came, Natalie would sit the girls down to manage their hair for the week. Natalie always managed Nadia's hair first, making sure she gave Nadia's hair extra attention, pressing out her coarse sandy brown hair and always adding bows or whatever it took to make her look extra special.

Nadia always cried from the heat that it took to smooth out the rough edges of her hair. Meanwhile, Christiana was always sitting on the sideline watching her mother manage Nadia's hair. Christiana wanted to get her hair pressed out like she had seen Natalie doing her sister's hair every time. She was forever asking her mother, "Do mine."

Natalie hostilely would tell Christiana, "Your hair does not need it," with an agitated tone. But from time to time, depending on her mood, Natalie would appease Christiana's request. Natalie would run the pressing comb through her hair for a minute just so that she would stop asking her.

Christiana would be content once her mother appeased her request, but she never understood why she did not need to have her hair pressed out. She was a tender-headed child. Her hair was soft and of a curly texture, which was easily managed. The one thing she hated was the brush because the bristles of the brush would penetrate right through to her scalp, scraping it after each rough stroke of her mother's heavy hand, which always caused her to draw up and pull away.

Christiana's tender headiness always had Natalie quick on the draw, jerking her daughter by her arms, lifting her small body from the chair at the same time, and then telling her to shut up because she was not hurting her. Each time she had to comb her hair, Natalie never missed the opportunity to push her in her back and then tell her, "Now go on and get off me!" Those comments always made Christiana feel deeply saddened and rejected by her mother, between the rough treatments and the rough brushings that always left her scalp burning, leaving it sore. The rough behavior from her mother always broke her heart.

Christiana started noticing that she hadn't ever seen her mother treat Nadia roughly, not once while doing her hair. These had become identifying moments for Christiana, where she realized that she was being treated differently from both of her siblings from an early age.

If Natalie was not at work or cleaning the house, she was then on the phone, mainly talking to her sisters. Christiana listened as her mother talked on the phone, hearing how she always praised her brother and sister, Andre and Nadia.

Christiana would listen, waiting to hear what her mother was going to say about her, but all she ever heard was a negative report. The things she heard her mother saying about her always ended with, "Girl, that damn Christiana cannot do shit right with her lazy ass!"

Christiana, by now, had become accustomed to hearing negative comments about herself from her mother, and even though she was used to hearing the negative comments, it still cut a new feeling.

Christiana had finally started elementary school. She was a quiet, easygoing student that caused no problems. However, she seemed to struggle with her classroom work no matter how hard she tried. Christiana had become so frustrated that she seemed to not be able to learn or keep up as fast as most of the children in her class. It was something that Andre and Nadia seemed to enjoy, and they teased her about it.

Christiana made it to the second grade, and she was still struggling. One afternoon the teacher became so upset with her as she kept asking the teacher for help over something that she had problems grasping.

This teacher got so frustrated with her. She then took her index finger and pointed directly at Christiana's forehead. She said, in front of the entire classroom, "Are you listening with your bird brain?" This comment humiliated her terribly. This teacher's actions left Christiana feeling humiliated and stupid, which brought her tears in front of the entire class.

Christiana arrived home to find that her mother had stayed home from work. She told her mother what the teacher had said and done. Natalie went up to the school to speak with the teacher about her conduct earlier the following morning. For once, Christiana felt like her mother cared and had her back, but that, too, was short-lived.

The teacher told Natalie that Christiana had taken things out of context and that she would never have said anything like that to a student. Natalie left the school and then returned home, but unknowingly to Christiana, Andre and Nadia had already been told about the comment the teacher made towards Christiana.

This comment was going to haunt Christiana throughout her childhood. It became something that both Andre and Nadia would say just to make Christiana feel even less during any arguments. Andre and Nadia both, at times, would chant towards her as they laughed, saying, "Shut up, bird brain," over and over until they felt vindicated. This empathetic taunting made their sister feel horrible, but when she would tell her mother what they were saying to her, all Natalie did was tell her, "Girl, if you don't go on with that noise, I don't want to hear that shit!

Christiana felt that she had no one she could confide in. She thought that maybe her father was the one that she could possibly turn to. Christiana was always looking to see who she might be able to trust or who might care to hear her plea for understanding. Christiana did not recall her father treating her harshly like her mother had. She also never heard him speaking negatively about her or directly to anyone.

To a huge surprise, one day, her father came and sat her down. He asked her how she might feel about living out of town, going to Arizona with one of her Uncles and Auntie. Christiana did not know what to think or how to feel about his question, but the first thing that came to her mind was fear. Christiana knew very little about either of her Aunt and Uncle other than their names. Besides that, how could she? Christiana was only eight years old.

Bash told his daughter that it would be best for her because she was struggling with her schoolwork. Her father went on telling her that they would be able to help her because they were both educators, but again, she would have to move out of state with them. It was at that moment she felt like her dad was throwing her away, which gave her more of a complex regarding her learning capabilities. Christiana didn't understand why it was so hard for her to learn. She told her father that she did not want to move with her Aunt and Uncle and that she would try to do better with her learning. What she did know was that she was falling further behind, mainly in math, because it was a very tough subject for her. These teachers continuously passed her on to the next grade with no concerns for her education; they were just getting students in and out.

Bash had some friends that he had rekindled with once he and Gia had moved back to Chicago. These friends knew his older siblings from when they first lived in Chicago. They always kept in contact with each other even after they had moved back to Louisiana after the death of their parents. Janice and Vincent Howard had moved to Michigan a couple of years after the two Bouvier siblings had moved back to Chicago. These friends, unbelievably, Natalie liked, therefore, their visits were always welcomed to their home because she could try to fool them about who she really was.

The Howards were coming for a week to visit the Bouvier family; they had discussed the visit on the phone while listening to the praise reports that Natalie always gave. Janice heard that Christiana was struggling in school with math, and once they arrived in town, Janice sat down to talk with Christiana. She wanted to find out exactly where she was struggling.

Janice spoke to Bash after the conversation with Christiana. She wanted to know about what they experienced when they were trying to help her. Bash told Janice that Natalie got frustrated and yelled at her, and they were just stomped with it to the point that they were considering sending her to live with one of his brothers and his wife.

Janice told Bash, as calmly as she knew how, that yelling at her was probably the biggest problem. Janice said, "I could see how she started to clam up when I started to speak with her about her school struggles." She also told Bash, "You and Natalie both need to learn some patience. That's all Christiana needs." Janice said, "Shit out of total frustration because of her parent's behavior!"

Janice helped Christiana practice her math the entire week and kept it at a pace for her to understand. Janice also allowed Christiana to keep asking her questions until she understood exactly what she needed to do. Christiana began to get some of the math problems, but she hated that Janice would be leaving to return to Michigan. Christiana wished she could be a good student someday with the hope that her parents might be proud of her, too.

Christiana was going to eventually learn how smart she was, but she just learned differently.

Chapter 20: Try Me

Christiana was growing older. She kept improving herself and was coming into her very own by no one else's standards. Christiana was about to move on to middle school, where she had planned on involving herself in after-school activities like volleyball and cheerleading. It seemed that Christiana was good at whatever she took a liking to.

Christiana had not yet realized that it was due to the ostracizing which unknowingly caused her to become somewhat of a perfectionist. The first day of school was about to begin, and she was ready for the new chapter of the middle school. She was excited to be moving on with all the junior high kids. On the first day of school, she entered the building and started to walk the halls when she noticed some of the girls were looking at her with glaring stares on their faces.

Christiana saw envy and dislike on some of the girls' faces as they were lined in the hall. She also heard this one group of girls saying out loud, "Oh, that bitch thinks she's cute." Oddly, she never viewed herself as being beautiful. What she did know was that these girls were being mean to her for no reason. Christiana was trying to figure out why she was being treated meanly because she knew that she didn't know any of them. She even tried being nice to the girls, but after a few failed attempts, she gave up.

Christiana learned quickly how to look for the signs of dislike. She was also able to see who was going to be a thorn in her side just by the looks the mean girls gave her, which resembled the treatment she received at home.

The Bouvier's other children were also moving into different areas of their lives. Andre was now fifteen years old, Nadia was twelve years old, and Christiana was eleven years old. Now that the children were getting involved in more activities, Bash insisted that Natalie needed to take a different shift. Also, Nadia had started having what her father thought was some form of regression and was displaying behavior that wasn't normal for a child of that age. However, Natalie still refused to get off the second shift. This left Bash with rearing the children all alone, and the children had become accustomed to only seeing their mother on the weekends. Andre was to get his sisters in after school until their father arrived home.

The children were only allowed to go no further than the front yard or the sidewalk unless Tammy or Rena said they could come over to their houses. Tammy always kept an eye on the girls to make sure that Andre didn't get agitated with them. Tammy could see how Andre wanted to hang with his older friends, and at times, he would holler at his sisters out of frustration for not being able to run with the older boys.

Tammy was even more concerned for Christiana because Andre had slapped her hand down on a hot burner one day, and another day, he had split open one of her toes with a fork tooth garden tool. Tammy would let the girls sit on her porch to play most of the time where she and her husband always kept snacks for when the kids came over to their home.

Tammy was always extra glad to see Christiana. Once she went over to the house, she made sure she hugged her a little longer so that she would know that she was loved by her. Tammy could see that Christiana knew how she felt for her and that she trusted her like a mother.

Andre's freshman year of high school had already started, and he began to slip out of the house at night after Bash fell asleep. He was sneaking out and going to parties with some of the older guys. Andre had also begun stealing his father's supply of alcohol that he kept in his well-stocked liquor cabinet, which Bash and Natalie used to entertain parties at their home.

Andre also started stealing his father's cigarettes and was on a path of destruction as he was trying to be grown way too fast. Bash was always getting on Andre's behind regarding his behavior, but Natalie would run straight to Andre's rescue, defending him from his father.

Natalie kept telling Bash that he was being too tough on Andre, but his mother was the one who bought everything that her son could think to ask her for. Natalie made sure he had all the clothes, electronics, and music equipment. The Bouvier's son had so much stuff that it was the reason that he had most of his friends. Bash kept telling Natalie, "You have spoiled that boy, and he is not going to be worth anything if you don't stop!" Natalie constantly ignored what her husband kept saying regarding spoiling Andre.

Bash was the first one in his household whose feet hit the floor at 4:45 a.m. Monday through Friday, his routine consisted of him checking in on his children first before he did anything else, making sure that all was well. One morning, when Bash took a bed check, he noticed that Andre was missing.

Bash woke Natalie up and told her that Andre was not in his bed or the house. Bash and Natalie's minds began to race with fear until a couple of the neighboring parents began to call each other, inquiring about their son's whereabouts. That was when Bash said, "Shit, I bet you that his ass is locked up."

Bash told Natalie, "I tell you, what if he's locked up? Believe me, when I say he is going to sit his ass right there for a couple of days because I'm sick of his shit, and Andre needs to learn a lesson."

Natalie began to cry, begging Bash to please not leave their son in there. She said, "Because he is my baby," she then said to Bash, "If you don't go get him, then I will go get him by myself!" Bash told Natalie, "That is what his problem is right now. You are always running to save his spoiled ass!"

Bash again told Natalie, "You are going to keep on, and he is not going to amount to SHIT. Do you hear me?" Natalie rushed to get dressed so she could go see if Andre was at the juvenile center. Bash then picked up the phone to call the juvenile center and to find out if they had Andre.

The juvenile detention center informed Bash that they had picked up their son and the other boys around 2:00 a.m. and that they were walking home from a party, also mentioning that he was intoxicated. They also informed Bash that they were about to start calling everyone's parents and that they needed to come to pick Andre up.

Once Bash and Natalie picked up their son and returned home, Bash told Andre that he was getting ready to go to work, but it was not over just like that and that when he got home later that evening, he was going to kick his ass!

Natalie abruptly stepped in and said, "OK, Bash, that is enough!" Andre was not even worried because he knew his mother would save him from his father's wrath. Andre then took his drunken butt to bed like it was the next best thing to do. A few hours later, Bash arrived home from work and was ready to deal with Andre.

Bash first noticed that Natalie had stayed home from work. She quickly stopped him before he could say anything to Andre. Once her husband walked in, she sheepishly said, "I have already talked to Andre." Bash said, "I told you that I was going to deal with him when I got home." But his wife said, "You don't need to say anything to him because he has already heard enough today." Bash just looked at Natalie, then shook his head as he walked off. However, Andre made sure to stay clear of his father for the rest of the week.

Nadia was still displaying some unusual behaviors. Bash felt it could be a regression or maybe from just missing her mother. Nadia had developed a problem where she was releasing her bowls in her underpants, then she would hide her underpants so that her parents would not see what she had done, but the smell of fecal matter always told on her.

Bash had to always ask, "Nadia, where are they?" Bash would then say, "And don't lie because I smell the shit." She always immediately began crying while pulling her shitty underpants up from wherever she thought was the next best hiding place. Bash was sick and tired of the hide-and-seek show with Nadia.

Bash would immediately make Nadia stand and wash the clumps of shit out of her underpants in the toilet. Bash was trying to make sense of her actions. He kept asking her what was wrong and why she kept doing that.

Nadia just continued to cry as she shrugged her shoulders. In other words, she did not know why. The behavior had gotten so bad with her crapping on herself that Bash had to come up with something that he thought would stop Nadia from shitting on herself!

Bash went out into his garage, and the kids could hear him throwing things around once he was inside before he found a 40-gallon bucket. He then placed it on the back porch with soapy bleach water. Bash called Nadia to come to the back porch and explained to her that it was for every time she took a shit in her pants, that it was where she was going to wash out her shitty underclothes! Bash figured that if she thought people saw what she was doing, it might break the problem of her shitting in her pants. Well, to Bash's surprise, the strategy had quickly worked and stopped Nadia's shitty behavior after about three buckets of porch washings.

Bash finally understood what was causing Nadia to contract what was known as pinworms. She contracted them from not washing her hands after wiping her behind after her bowel movements. Bash figured out that she had worms once she kept digging in her behind due to the pinworms crawling in and out of her rectum, which caused her butt to itch.

Nadia would then touch things in her and Christiana's room, and shortly afterwards, her sister would also end up with the pinworms. Tammy told Natalie to get rid of the pinworms and that she needed to give the girl a tablespoon of turpentine with sugar, which would rid the girls of the pinworms. Christiana made it a point to take a spoonful each week afterwards to safeguard herself because she never knew what to expect.

One evening, Christiana was outside riding a bike that was way too big for her, and it also happened to be a boy's bike. She slipped and fell, hitting her vagina on the bike's bar very hard. Christiana ran into the house due to the pain and wetness she felt between her legs. She went into the bathroom and pulled down her underpants, and all she saw was red. She realized that she was bleeding.

Christiana hadn't yet started her period, and she knew nothing about a period either. She went into where her father was sleeping and woke him up, showing him her blood-soaked underpants. As Bash opened his eyes to see what his daughter wanted, he looked down at what she was showing him. Bash looked wide-eyed and then rolled out of the bed onto the floor. He jumped and then ran to get his neighbor Tammy because he did not know what to do.

Tammy came running to Christiana's rescue. She asked her, "What happened, baby? Let Tammy see?" Tammy opened her legs to look at her vagina to see if she was injured or if she needed to seek medical attention. Tammy told Bash, "I am not sure what to think, but I think you should take her to the emergency room."

Bash rushed and took his daughter to the hospital. The doctors examined her and then told him that she was going to be fine and that she had torn her vulva from the hit on the bike's bar. The ER doctor told Bash that he could take his daughter home and to apply the ointment he was going to prescribe and make sure to let her rest so she would heal.

Natalie arrived home from work around midnight. That night, she found Bash waiting to talk to her. That was when she found out what had happened to their daughter. Bash, without any hesitation, said, "Natalie, you need to be home with these girls because this is exactly what I've been talking about that could happen! All I know is that had it not been for Tammy, I would have not known what to do! Natalie, it was a lot of blood, and besides that, the girls need you here with them."

The next day, Natalie got on the phone, telling Mr. and Mrs. Montgomery her version of a lie that she had created about what happened with Christiana, stating that she had let some boy mess with her sexually, which caused the injury. Natalie's lie regarding her daughter led Mr. Montgomery, Christiana's grandfather, to tell one of his other daughters to keep her child away from Christiana because she was nothing but trouble.

Natalie again refused to take another shift, but the real reason was that she was attracted to one of the guys on her shift. One of the men had even become her rider for work. Another one of the men was a married man with whom she had developed a more intimate relationship. His name was Edward, but that relationship would soon lead to something more than Natalie could not imagine.

Chapter 21: Sidepieces

Bash and Natalie were always having parties at their home. At one particular party, Natalie invited Edward and his wife Sandra to their home. That was how careless she had become. Natalie always thought she could pull the wool over her husband's eyes, but not everyone was blind. Edward's wife, Sandra, and Edward's sister, Gloria, came to the party at Natalie's and Bash's house that night.

Bash was always a very pleasant, perfect host to his guests. He welcomed Edward, his wife, and his sister into their home. After a couple of weeks, Edward and Sandra invited Natalie and Bash to their home for a party. They even took Nadia and Christiana to play with Edward and Sandra's daughter, Camille.

This partying group was about to get even more interesting. Edward's wife, Sandra, figured out that Edward and Natalie were more than just casual co-workers. Sandra noticed that they kept watching each other, and she decided that she was about to level the marital playing field between Natalie and herself.

Sandra set her attention on Bash so that the game would be more evenly interesting. Sandra was a 5-foot-5-inch curvy and good-looking woman with very fair flawless skin, who wore a skintight fitting dress to show off her vivacious figure and she was going to try to entice Bash's appetite after she realized that there was more going on with Edward and Natalie.

Sandra felt confident that she could get Bash to become more than a willing participant with just a little coaching. Natalie overheard Bash on the phone one night, talking to Sandra. Natalie instantly jumped to her conclusion and then began to accuse Bash of being slick with Sandra despite her and Edward's extramarital relationship.

Bash just laughed at her and told her that she was taking things way out of context. However, he enjoyed seeing Natalie being blatantly jealous for a change. Bash said, "Natalie, Sandra only called to inform us about a surprise birthday party that she is going to be throwing for Edward," but Natalie still was not buying it.

Natalie and Bash went to the party that Sandra was throwing, and the entire time they were there, Natalie kept a watch on Sandra and Bash. She wanted to see if they were going to make any kind of eye contact so that she could know what her motive was towards Bash.

Sandra made sure she catered towards Bash, and Natalie was taking a shitty notice of her attempts. On their way back home, Natalie said to Bash, "I see Sandra's ass was making sure that you had everything that you needed."

Natalie couldn't even contain her anger and went on to say, "That bitch did not one time ask me if there was anything that I needed." Bash responded, "Well, it probably was due to Edward, who was seemingly making sure he was taking care of your every need." Natalie asked, "And what in the fuck is that supposed to mean?"

Bash laughed and said, "Just what I said, I'm pretty certain that I didn't stutter. Natalie, Edward seems to know exactly what you drink, what you eat, and damn near the color of your fucking drawls. Now, isn't that amazing how he could know so much about who is supposed to be just a coworker?" He snickered. Natalie became furious, then sat and remained quiet the rest of the way home. She was only building up more ammo for once they arrived home.

Once they arrived home, Nadia and Christiana went straight to bed but were unable to get any sleep because Natalie had restarted the argument with Bash all over again. Natalie said, "That bitch better not ever call this house again and let me catch you talking to the bitch, then I'm going to kick your ass too!" Natalie still didn't let up. She then said, "Just know that I saw you skinning and grinning at that bitch, which I know you didn't think I saw but fuck you and that bitch!"

Bash again laughed, which was pissing his wife off further, while he just kept agreeing with his wife by saying, "Ok Natalie, whatever you say, baby," still laughing. The argument was tickling Christiana because she understood that her father was winning an argument with classic sarcasm without verbally attacking their mother.

Bash finalized the argument once he questioned his viewpoint of the night. He said, "Natalie, I'm really trying to figure out what in the hell you are so damn mad at her for. But here's a thought - maybe you should ask Edward. I think you should ask Edward to see what he thinks about it or what he has to say." That comment shut Natalie's mouth up immediately. Bash then doused flames on her argument and asked, "What did you say because I can't hear you?"

The next morning, Natalie called to talk with her sister, Adrian, to give her the rundown of the party that she and Bash had gone to the night before. Natalie said, "Adrian, girl, let me tell you about that bitch, Sandra." Her sister asked, "Who is Sandra?" Natalie replied, "It's Edward's wife with her funky ass!"

"That strumpet damn near did everything other than fuck Bash while he sat there skinning and grinning with her slithery tail!" Adrian asked, "So where was Edward, or what did he have to say?" Natalie said, "He was there, but get this, Bash had the nerve to say something about Edward and me on the ride back home after I jumped in his ass about Sandra. I tell you one thing: she better not ever call this house again." Adrian said, "But wait, Natalie, aren't you and Edward involved? So why are you so mad?"

Natalie said, "Yes, we are, but that doesn't have shit to do with nothing. If she thinks she's going to just do some disrespectful shit directly in my face, with Bash sitting there looking like a damn cheesing chest cat in the fucking face, she has another thing coming!"

Adrian asked, "Natalie, aren't you doing the same thing to Bash? So what's the problem?" Natalie said, "No, it has been done differently!" Adrian said, "I really don't see how it is, but whatever you need to tell yourself because creeping is cheating!"

Chapter 22: Party Fillers

Natalie made her mind up that she needed to introduce her sister Adrian to one of Edward's friends to aid her in her scheme of spending more time with Edward. Natalie was always crafty when thinking about how she would be able to exit the house easily with the assistance of Adrian, realizing that Bash would not think very much of it if her sister was with her.

There were many times that Natalie would pick her sister up and then come back to the house just to sit for a short period before she would make up an excuse that they were going shopping. Natalie, at times, even took her daughters as part of her diversion but then would end up pulling up to some strange man's house whom the girls did not recognize.

When these strange men would come out to speak with Natalie and their Aunt Adrian, Christiana noticed that the men were more interested in only talking to their mother rather than their Aunt, who was a single woman. Natalie would make sure to make it clear to her daughters that they better not say anything. A threat her daughters didn't take lightly.

Natalie would only hang with people who would help her by scheming lies to fool her husband. Now, if they would not help her scheme, then Natalie had no use for them. Natalie one day tried her neighbor Tammy to see if she wanted to meet someone. Meanwhile, Tammy told her that she was not interested in meeting any other man. Tammy also told Natalie that she already has a husband who is a good man, and so does she, but that was her business with what she does with her marriage.

Natalie didn't like how Tammy made her feel, and suddenly, she developed a problem with her. Tammy was not concerned about how it made her feel but was shocked that she came to her for an extramarital affair. However, she continued her relationship with Christiana even as she grew older. Their relationship grew even stronger. Tammy knew that later in her life, Christiana would learn just how close they really were.

Christian had turned 14 and was understanding more than she should have known by that age. She understood that her mother was one lying, cheating woman, and she could not understand how her father was able to stay with her mother. Christiana understood her father valued his family, and that must be the reason why he had not left Natalie. Christiana realized that a person would have to be deaf, dumb, and stupid to not see through that type of behavior. But little did he know, he was teaching his children to accept and to stay in some junk while they were trying to create a promise that was never meant for them.

Christiana had a lot of respect for her father for many reasons. He made sure that he and his children sat down at dinner time together to eat each day. Bash also made sure to find out what was going on with their lives, and once they were done eating, he always had some project that he had created and was working on, but he made sure that he included his children in what he was working on; this was always a delight for them all.

Bash made it a point to teach his children a lot of things about life and how to take care of themselves with many skills. Christiana learned how to check the oil in a car, she learned how to cut grass, she learned how to listen to gain knowledge. She even learned how to lay sealant on a black top. She learned how to take care of children, and she learned about sacrifice, all from her father. Bash made sure to let them see how to look out for the neighbor's kids who might be in need and he was forever engaging with the neighborhood kids who also had the utmost respect for him.

The one thing that Bash never realized was that Christiana paid closer attention and that she realized that his feet were the first that hit the floor first thing in the morning while going to make a living for his family. Bash made sure that his children did not know what it was like to go and peek into a barren refrigerator. He also made sure that their needs were always met constantly.

Christiana made it to the 8th grade. Her siblings were both in high school. Natalie still stayed on the second shift, and Bash had given up on asking his wife to take another shift. The Bouvier children had grown and adjusted to the schedule of their mother not being there during the week.

Chapter 23: How It Feels

Christiana's struggles had changed from schoolwork problems into another struggle of feeling like an outsider with her family. One evening, Natalie, Bash, and another couple had gone out to a supper club. There was some woman who must have said something to Bash in possibly a flirtatious way because things got heated.

After about an hour, both couples arrived back home. There was a heated argument that took place between both couples. Christiana was already in the kitchen when they came in, and all she heard was cussing and fussing.

Bash raised his voice and said, "I don't give a fuck about none of yawl!" Christiana said, "Don't say that, Daddy!" That's when Bash said to Christiana, "You shut the fuck up because I remember that there was a time when we didn't even want you!"

The room fell silent. Christiana ducked her head and then left the room, but not before seeing her other two siblings discussing what they had just heard their father say to their sister. After that comment, no one came to see if Christiana was ok after her father's comment, nor did Bash ever come to apologize to her for what he said.

Tammy was always watching Christiana if she happened to be sitting out on the front porch. Tammy would come out on her porch just to talk to Christiana from across the street. Christiana would always get up to go over to her and Reed's house because she always knew that

Christiana would come just to talk to her about anything, if nothing, other than her flower garden.

One day, when Tammy and Christiana were at her home sitting at the dining room table, Tammy felt Christiana was feeling sad. Tammy told Christiana that she paid more attention than she would believe. Christiana realized, at that moment, that what she was feeling had not been just her imagination. Tammy said, "Let me tell you something. You have always been my baby, and you will always be my baby," and Christiana knew it.

Tammy also said, "I can hardly stand that damn Nadia with her smartass mouth, and it's your mother who approves of Nadia's nasty behavior." Tammy also said, "Nadia thinks that her ass is grown or on the same level as an adult, but again, Natalie has told her that." Tammy went on to say that Nadia and Natalie didn't realize that I had damn near smacked the shit out of Nadia one day when she decided to make a negative comment regarding one of my grown sons, and Natalie did not open her mouth to correct her behind. I left before I would have whipped her ass right in front of her mother.

Tammy said, "There's one more thing that I'm going to tell you, and you are going to see it for yourself one day. The Most High is going to take your father first because your mother has a lot to atone for, and I mean a lot! Christiana, baby, I see how those two are always jumping in the car, leaving together, and always leaving you behind. Just know that Reed and I are always here for you, and I love you like you are mine."

Tammy continued talking about Natalie, then said, "The real reason that your mother does not deal with me anymore is that she wanted me to help her to be able to cheat on your dad. I could not believe that Natalie offered me to meet one of her guy's friends so it would look like we were going shopping or some old crazy shit."

Christiana felt mentally relieved after listening but still felt like it was all in her head that she was being treated indifferently, regardless of what she was told, which still hurt her heart. Christiana could not make sense of the reason why there was a difference, but she would soon come to find out the truth one day.

Andre was on a downward spiral, and he would learn that he would not be able to control it. He had started dating his high school love, and he also had an opportunity to attend a big college, but he chose to stay on the path of wanting to be an adult with all the responsibilities that came with it.

Andre went to his parents one day to inform them that he had gotten his girlfriend Lauren pregnant. Bash gave Andre the option that he and his mother would take care of his child financially until he was to graduate from college.

Natalie was upset about the pregnancy while telling everyone that Lauren had gotten pregnant on purpose just to trap her son. Andre chose not to attend college, stating that he was a man and that he would take care of his own child.

Bash and Natalie went to secure them an apartment, then completely furnished it with every household item for a start, and tried to set them up with a good environment. However, Andre had already started drinking heavily and staying out at all times of the night. Andre was also keeping Natalie abreast of each detailed argument between him and his girlfriend, Lauren.

Natalie had developed a dislike for Lauren before the baby was even born. She blamed Lauren for all the fighting and arguing concerning Andre. Bash finally told Natalie that she needed to stay out of their son's business because Andre had stated he was a man and that she was aiding him in his nasty ways.

Natalie was livid with Bash, as usual, when it came to him telling her about herself and their son. Andre's son Manias arrived, and Andre felt like he was the man. One evening, he took the baby over to his parent's house along with Lauren for a visit.

Christiana was excited once they arrived; she went to hold the baby, and through her excitement, she said, "Hey, little stinky butt." That was when Andre went off on Christiana. Her brother stated, "Don't start that shit. You are to call him Manias!" Christiana immediately handed her nephew back to Lauren and then left the room while thinking to herself, what an ass!

Natalie was always shopping, buying her first grandchild everything that she could think of because she wanted him well-kept and dressed as she had done with Andre. Andre's girlfriend ended up nicknaming their son 'Cheeks,' which Andre and the entire family ended up calling him by. Andre and Lauren ended up getting pregnant with another baby when Manias was seven months old. They ended up with another son who they named Ya'fam.

Christiana became a built-in babysitter, but it didn't matter to her because she loved her nephews, and she enjoyed teaching them new, fun, silly things, at times just to agitate Bash, their grandfather. One visit, while Bash was sleeping, Christiana had gotten Manias to take a stuffed animal and wake up his grandfather. Once Bash opened his eyes, the toddler was standing there. He then pointed under the animal's tail and said, "Doo kee shoot!" Bash called out from being shocked. He then said, "Alright, Christiana, I know I didn't anyone teach him that but you," as he chuckled.

Andre and Lauren were constantly arguing and fighting. Natalie always had to add her spin on their arguments and would talk about Lauren to her son. One day, Natalie told Andre that heifer's kitchen on the back of her head looks like fucking pubic hair on her neck and that she needs to clean herself up.

Natalie then said, "Hell, I went into the bedroom and saw her drawls lying on the floor this morning, and it looked like gum was in the crouch of her drawls." Natalie said, "I had Christiana bring them nasty-ass drawls in here to the washing machine because I wasn't going to touch them."

Chapter 24: Medical Urgency

Christiana had developed a bleeding problem that carried on for about a full month. Once she was taken to the hospital, they decided that she needed to be hospitalized to run tests to find out what was going on with her health. Christiana was hospitalized for roughly a week before it was determined that she had a cyst on one of her ovaries that had ruptured, which caused a lot of the bleeding.

Natalie had gotten on the phone and spread a rumor that Christiana had gone and had a back alley abortion while stating that is what was really going on with her. Christiana was informed by Lauren of what her mother was telling everyone. Christiana told Lauren that she did not know why she would be telling anyone that, knowing that her mother knew the truth based on what the doctor had informed her parents about what had taken place.

Chapter 25: Matters of The Heart

During Nadia's freshman year of high school, there was one young man who caught her attention, and she was interested in him. So, one night, Nadia snuck out of the house once her father had gone to sleep, but she did not make it back in time before her mother arrived back home from work at midnight.

Natalie arrived home and then realized that Nadia wasn't in bed or the house, once she went and checked on the children. She woke up Bash terrified about where Nadia was at. Natalie and Bash had gotten in their car and began riding around the neighborhood, looking for her. Eventually, they found Nadia walking back home, coming from the park where she had been with the young man whom she liked from school. They had been in the park in one of the dugouts, making out from the time she had snuck out of the house.

Nadia was very much like her mother when it came to sneaking out of the house or not being where she said she was going to be. Nadia was forever crying once she got caught and shrugging her shoulders when it came time for her to explain her actions. Natalie always kept Nadia's secrets quiet, but that was going to change soon.

A year later, Christiana started high school, and without being asked, she was deemed one of the most popular and beautiful girls on the school's campus. Christiana was much like her grandmother Madeleine, who also had a sweet spirit that drew people to her.

In Christiana's freshman year, she was approached by an upperclassman who talked to her daily in passing through the halls at school.

Christiana was one easy person to talk with because she made conversating pleasant, regardless of who you might have thought you were or even if you thought you weren't. The upper-classman stopped her in the hallway between classes one day and asked if she would be his date for his senior prom. He acted surprised once he saw "Christiana" not having any reservations. She did not hesitate to accept, going to the prom with him.

Christiana had taken notice of a young man the summer before her sophomore year without him even knowing it. The young man, Kyrie, seemed shy and quiet, but that was what attracted him to her. Christiana did not know that they would end up having summer school classes together the following year. She was excited to see that they had a class together with hopes of getting to know one another.

Kyrie immediately began walking her home each day after school. He then stayed with her until late in the evening. Christiana had become curious as to why he stayed with her each day. She made up her mind to ask him about his extended visits with her. Christiana went to walk him out for the evening, then asked him, "I need to know why is it that each day you seem to keep hanging around me. Is that you like me or what?"

Christiana, deep down, was hoping that his response was going to be yes, and to her surprise, Kyrie admitted that he did have a strong interest in her. Christiana said, "Why didn't you just say that? I didn't understand why you keep hanging around me?" But she also saw that he was relieved that she had asked him because he had not built up the courage to let her know that he was interested in her. Now, just minutes before Kyrie was about to leave, and right after their conversation, there came this one guy named Lenox, who was also a neighborhood associate who also liked and hoped to be Christiana's boyfriend! Lenox had been trying to call her to get a date, but she wasn't interested in him. Therefore, she did not take any of his calls.

Once Christiana didn't give Lenox the attention that he so desired, he decided to drive past her home and pulled up in front of her home. He saw Kyrie standing there with Christiana. Lenox hopped out of his car with an attitude, like he thought he had a right to be there, and began to question Christiana.

Right away, Lenox said, "I have been trying to call you!" Christiana was pissed that he had the gall to even try and question her. Without haste, Christiana said, "I know," and then waited for his next response. Lenox's emotions had gotten the best of him once he saw Kyrie with Christiana. Neither of them cared about him being there.

Unbelievably, Lenox asked, "So, is that your man?" Christiana then looked at Kyrie to see what his response was going to be to the question. Kyrie pulled Christiana back towards him, and that was when she looked back at Lenox and said, "He is." Meanwhile, Kyrie never let go of Christiana to confirm for both Lenox and Christiana's questions. As soon as Lenox got the response, he angrily hopped back into his vehicle and drove away. Kyrie pulled Christiana closer to him and then kissed her, and laid his stakes on her from that moment.

Kyrie treated her with love, but what he did not know was that it was the first time that she had ever been made to feel loved by anyone. Christiana fell for Kyrie because his love for her appeared to be consistent seeing how he always came for her.

Christiana believed she was able to let her guard down with him. She also believed that she could confide in him with things that she would not have otherwise shared with anyone else. The following year, Kyrie gave Christiana his first promise ring, which she happily accepted.

Natalie could not stand the fact that Kyrie was making plans with Christiana because she could see how well they were bonded. Kyrie and Christiana both talked about marriage and the dreams of having children together one day.

Everyone thought that they made such a beautiful couple due to the genuine way they seemed when they were together. Natalie could not hold in her dislike for Kyrie, so one day, she stated to Christiana that she found him to be arrogant.

Christiana knew that her mother's real dislike for him was not based on his attitude but more on how well he treated her daughter and that he had never given reason for her mother to believe otherwise. Now Christiana knew that if it had been Nadia who was dating him, Natalie would not have had any issues with him at all.

Christiana and Kyrie had an unspeakable bond. The one thing they really loved was taking long walks together because it was their bond builder. Their walks were much like Christiana's grandparents' Jacques and Madeleine's walks, where time held no measure once they were together.

Christiana loved and trusted Kyrie without any doubt until one evening, she went outside to take a short walk by herself. As she had gotten to the corner near her home, Christiana looked down the street and saw Kyrie driving some other female's car with her sitting in the passenger seat. The female just happened to be a friend of one of his sisters, but the one thing that Christiana knew was what cheating resembled.

Chapter 26: Them Drawls Hot

Natalie was out of complete control. Not only was she still messing around with Edward, she had found another man. She decided to mess around; however, this time, it was one of Bash's frenemies, Scott, who was pretending to be a friend to Bash while fucking his wife.

Natalie and Scott had been creeping around for months until they felt that they had to come up with a plan to buy the girls off for their deception. The plan was to give gifts to the girls who were nothing more than hush-money gifts. Scott had gone and purchased the girls some bikes. He told Bash that he had found the bikes on sale because he had noticed that the girls did not have any bikes, and that was why he grabbed them for the girls. However, the girls had not mentioned wanting bikes.

Natalie suddenly decided to take a medical leave from her job, claiming that she was under stress. Every morning after Bash left for work, around 45 minutes later, Scott would arrive at the house and then Natalie would go into the kitchen to prepare him breakfast that Bash had paid for.

Scott would always leave about an hour before. He knew that Bash should be arriving back home from work. Scott was always bold enough to come back, bringing cold beer for him and Bash to drink. Natalie would then go and prepare dinner to make sure that not only did Bash eat, but so did Scott.

The things about the Bouvier household were no longer private to just the family. It came to the point that the girls could not even get out of bed and come out of their rooms in their gowns and walk freely through the house if they chose to because Scott was always sitting around in the house like a fixture. One day, Christiana said to Nadia, "This bitch probably even knows when we are on our periods because our mother tells him everything that goes on in this house."

Scott didn't care for Christiana because he knew she had an outspoken personality, but more so when it came to speaking the truth. Her mother or her sister also didn't like that about her. Natalie and Nadia, at times, would try to demean Christiana because of her outspokenness, mainly if she spoke something that rang true. They were forever calling her Gia, referencing or comparing her to Bash's sister. They believed that it would affect her spirit while trying to break her. However, Christiana's spirit grew stronger with age. She realized that she was nothing like them; therefore, she took their comments as compliments because she also understood envy.

Natalie was forever picking arguments with Bash because she was trying to figure out her escape plan so she could date more freely. Natalie finally picked a certain day to make her exit. She then went and started an argument with Bash and told him that she was leaving him. Natalie told the girls to get some clothes together and then took them with her to her sister Adrian's house for the night while she figured out where she was going to house them all.

The next day, Natalie and Scott went looking for a place for her and the girls. She then found a duplex that they could move into right away. As soon as they moved into the duplex, the girls heard that rats were running in the walls. Natalie, after a couple of weeks, realized that it was the slums. She then hurried and moved them into an apartment that was just a step above for a while.

The new apartment that Natalie chose, the girls knew, was not a good or safe place for them, but they had to follow their mother. Natalie, out of fear, thought that Bash would question the girls and find out the truth about her affairs. That was why she would not leave them with their father.

Right before Christiana found out that she was still being cheated on by Kyrie, he made sure that he found and stayed in touch with her, regardless of where they moved to. Natalie once again moved them into another seedy apartment complex, but by then, the girls thought that their mother had lost her mind. Christiana immediately went and got a job just to be able to help with bills and groceries because she felt the struggle, and it was something that she was not used to, nor did she want to understand it.

Nadia had started dating a guy who was a relative of the neighbor who lived next door to Bash. It was one of Rena's nephews named Breon, who really liked Nadia, and she liked him, too.

Breon had begun spending a lot of quality time with Nadia, but Natalie didn't like the fact that she was dating Rena's nephew. Breon was crazy about Christiana, and she was crazy about him, too. They were forever laughing with each other, and she hated that her mother had a problem with Nadia and Breon's relationship because she knew that he really liked her sister.

Natalie noticed that Nadia had not asked for any sanitary items for the month shortly after they had begun dating. One weekday morning, right before they were to get up to get ready for school, Natalie angrily woke Christiana up from her sleep. Once Christiana opened her eyes, she noticed that Nadia was sitting on the side of her bed, crying. Her first thought was, Oh shit, this ain't good! Natalie, with a hostile tone, said to Christiana, "You both are not going to school this morning because I'm getting ready to take you all to the planned parenthood clinic!"

Christiana put the whole scenario together right away, with Nadia crying and Natalie mad, so she hurried up, got dressed, and then waited for her mother and Scott, who was taking them to the planned parenthood clinic.

Once they arrived at the clinic and signed in, some nurses called both girls back at the same time but placed them into separate rooms. Christiana was examined, and her findings were that she was not pregnant. The nurse gave Christiana a bag of different types of birth control paraphernalia, and she was told that they would be letting her mother know how her exam went.

Christiana was not the least bit concerned with what her outcome would be. However, she knew that her sister's outcome was not going to be the same. The nurse walked Christiana back to the waiting area, where Natalie and Scott were seated. Christiana then strolled over, and as soon as she sat down, Natalie looked down her nose at Christiana and then aggressively said to her, "It's just a matter of time before you and Kyrie's shit comes out in the wash!"

Christiana's eyebrows raised as if to say wow, and she humorously thought to herself, I thought we were here for Nadia! Christiana also thought but didn't say it, that, no, it's just a matter of time before your shit comes out in the wash!

Natalie was then called back by one of the other nurses concerning Nadia. Once they returned to the waiting area, Christiana saw that Nadia did not have a sack of birth control paraphernalia. Christiana damn near wanted to chuckle at her mother.

Natalie looked heated while Nadia was snotting as usual. Christiana remained quiet on the ride because no one was saying anything. Christiana was taken and dropped off at school, then was told that her father would be picking her up later that afternoon.

Bash arrived to pick Christiana up later that afternoon. Once Christiana got into the vehicle, she could already tell that Bash knew that Nadia was pregnant. Bash suggestively questioned her by saying, "Well, I take it that you've already heard about your sister, right?"

Christiana felt like it was a trick question, so she just nodded her head to let her father know that she knew. That was when Bash said to Christiana, "You know that it is your fault, right?"

Christiana was stunned by her father's accusation. Christiana just stared out of the passenger side window with bucked eyes, thinking to herself, now, how in the fuck is this my fault? Christiana badly wanted to say, 'Hell, I didn't hold her damn legs open, nor did I say, 'Oh! You're fucking today, ma'am!'"

Christiana had learned to let whatever accusations they blamed her for concerning everyone's mishap just be. Christiana knew she didn't need to defend her character because she knew that her hands were clean.

Natalie and Bash made sure, without discussion with Nadia or Breon, that she was going to get an abortion, all to save their image and Nadia's image; then, it was never spoken of again.

After that pregnancy with Nadia, all Christiana ever heard from the family were put-downs, where they complained about everything concerning her. They would make it a point to say things like your lips are too black, your legs are too skinny, your hair is too fuzzy, or you don't have any behind. They said anything to body shame her to feel good about themselves.

Christiana had begun to internalize all the put-downs, even though she was able to look in the mirror and see that she was attractive. They had made her feel like she was not good enough. Christiana knew that her mother and her sister were color-conscious even though she was a shade darker than either of them. Christiana understood and hoped that she would not have any light-skinned children because she was fearful that they could possibly have the same sick mindset, thinking that they were the cream of the crop only if they were lighter.

Chapter 27: Secret Combination

Kyrie had seen Natalie's indiscretions, but not once did he lay judgment towards Christiana's mother, even after he had seen the times before Natalie left Bash and the times after she had left Bash dealing with having different men.

The latest apartment that Natalie had moved into with the girls had five locks on the front door for security purposes to keep anyone from getting in on them from the outside. One day, Andre asked Kyrie if his mother was seeing anyone.

Kyrie did not know that it was to be kept a secret. He figured that it was no secret because Natalie had already left Bash. Kyrie told Andre that his mother was seeing a couple of men, but to what magnitude, he did not know.

Andre asked Kyrie who they were and what they looked like to make sure he gathered all the details before he made his presence known to them. One evening, the girls were home with some of their friends, including Nadia's new boyfriend, Vincent, and some of his cousins. One of Natalie's lovers was also present.

Edward and Natalie were in her bedroom with the door closed, not knowing this was the evening that Andre had been sitting outside waiting and watching to see who was going to show up to visit his mother.

Andre waited and let Edward get good and comfortable before he decided to go and knocked on the apartment door to gain entry. Once Andre got into the apartment, he then put all 5 of the locks on the front door. He made sure that Edward would not be able to make a fast run from it once he located him in the apartment.

Andre stood around talking to everyone that was in the living room for a few minutes. He then acted as if he needed to use the restroom. What Andre was checking for was Edward's hiding options so that once he located him that he had plans to whoop his ass.

Natalie heard Andre's voice and had Edward hide in her bedroom closet. Andre came back into the living room and talked for a few more minutes, but he was only making sure that the locks were still intact as he had first left them upon his entry.

Andre asked his sisters if their mother was there. They told him that she was in her bedroom. Andre went and knocked on his mother's bedroom door. Natalie was in a house gown when she came to answer Andre at her bedroom door.

Andre was shitty upon seeing his mother, but he played it calmly. He then walked in like he just wanted to talk to his mother, but he was just narrowing down Edward's hiding options. Andre did not see him in the bedroom as he was slowly moving about the room. Andre then thought, 'This motherfucker bet not have his ass in this closet.'

Natalie looked nervous and sheepish in the face, hoping that her son was going to leave, but he kept rambling like nothing was going on. Andre moved in a rush, then snatched the closet door wide open. He saw that Edward was hiding from him, and right then, Andre commenced dragging and beating Edward all about his head and face. The fight had made its way into the living room, where everyone was. Edward was trying to get away to save his behind from taking that ass whopping.

Natalie was screaming for Andre to stop as Edward was still trying to get to safety, but again, all five locks were still intact on the front door, which was one hell of an ordeal for Edward in making a fast exit. Natalie was trying to hurry to help get the door opened for Edward to escape. Edward and Natalie both were struggling with the locks while trying to keep Andre off him. Once they got the door opened, Edward didn't even take the time to run down the stairs. He decided to clear an entire flight of steps by jumping to the next landing to get away.

The house guests were looking in amazement, thinking how he should have broken an ankle or a leg with the jump he made, but they did not involve themselves to stop it either. They were enjoying how entertaining the fight was while laughing at Edward's scared face and how he could not get the door opened fast enough because of all the locks.

Vincent said to everyone, "It looked like Edward was working a Rubik's cube, trying to get the locks undone while trying to keep Andre off his ass, then he was forgetting which lock he had or had not touched!" The guest and the girls could not hold their laughter as Vincent kept going to the door, reenacting the door scene. They were all trying to muffle their laughter, but it was impossible until Natalie began her questioning.

Natalie was shamefully embarrassed as she kept asking Andre who told him about Edward. Andre, without any hesitation, simply replied that he asked Kyrie. Natalie immediately turned towards Christiana and then told her, "I don't want that asshole back over here since he's been running his damn mouth and starting shit!"

Christiana called Kyrie to ask if Andre had questioned him about their mother and her dating. Kyrie said that Andre asked him, but he did not know that it was supposed to be kept a secret. Christiana said her mother was blaming him for exposing the truth. Kyrie asked, "How is it that your mother wants to blame me for her crate of skeletons?" Christiana said, "She's always looking for a fall guy, and I am sorry; I just wish I could get away from them."

Chapter 28: Bailing Time

Shortly after the altercation with Andre, Christiana saw Kyrie with another female, which left her feeling disenchanted with him. She could not believe that he would cheat on her because she believed they shared a special bond.

Several months later, Kyrie joined the military, but before he left, he visited Christiana. He wanted to let her know that he had enlisted and would be leaving, but the sting of that hurt was still present because he never addressed the conversation with her.

Kyrie arrived at her home. He stood out on the porch to talk to Nadia and then maybe ask her to get Christiana for him, but before that could happen, Nadia went into the house and told her sister that Kyrie was there to see her. Christiana didn't know what to feel, but she picked up that first promise ring that he gave her and went out to find out what he wanted with her.

Christiana first handed him his ring back. Kyrie tried to make light of her actions by saying to Nadia and another friend who was on the porch listening to them, "Look at Christiana, she is mad at me," but he did not understand that it was hurt and disappointment due to his behavior.

Kyrie then went on to say, "I came to tell you that I'm about to leave to go into the military."

Christiana turned and went right back into the house because she did not know what he wanted her to say, except that it felt like abandonment all over again. Kyrie sent a few letters to Christiana once he was stationed. They did correspond with each other for a short while before the letters eventually dropped off.

One afternoon, as Christiana was sitting on the porch, she happened to look up and saw Kyrie coming down the street heading towards her. Once they locked eyes, all the old feelings came rushing back. Christiana could tell that he was glad to see her, and she was certain that he could tell that she was just as excited to see him. Christiana didn't want to part from their visit, but she had to go to work.

Kyrie asked, "Why don't we get together after you get off work this evening?" Christiana told him that it was good, but she wanted to change her attire first, and then she would come around to pick him up. Kyrie agreed that it did sound good to him and that he couldn't wait to see her.

Christiana got off of work. She then hurried home and changed into a nice dress. She then drove around to pick him up. He then took the lead and drove her vehicle as they headed out to a club just to dance. It felt like old times, as if time had never stopped between the two of them. After a few dances, they decided to go downtown to walk and talk because it was something that they both loved doing together and where they could disconnect from the world and enjoy their bond.

Kyrie kept singing her a song called 'Remind Me' as they were walking while Christiana listened to the message of what he wanted her to know. Kyrie kept hugging Christiana. He then pulled her in for a kiss. That kiss was one of those once-in-a-lifetime, unforgettable types of kisses she had to pull away from him because it made her lightheaded. Kyrie had to catch her to keep her from falling!

Christiana also knew that he felt exactly what she had felt because he would not let go of her. Christiana quickly stated that they should be headed back home because she had work in the morning and that she was also cold. Kyrie said, “You know what warms that up, right?”

Christiana didn’t ask what he meant by his comment, nor did she give a response to his question. Kyrie then stated that two warm, naked bodies would warm her up. Kyrie again waited for a response, but Christiana was not one to play games or assume anything because she remembered being disenchanted by him before.

They continued to walk and talk some more before he dropped a bomb on Christiana. Kyrie nervously said, “I am getting married next month.” Christiana’s heart sank, but she congratulated him out of courtesy. Christiana immediately wondered, ‘Then what in the hell is he doing here with me?’

For the next two weeks, Kyrie spent every day with Christiana as if he did not have some other female waiting on him. There was one night when they went strolling, their vibes mirroring each other, and they couldn’t help constantly embracing each other. Kyrie told Christiana that she could make him change his mind about getting married to his fiancé. Christiana chose not to say anything because she knew that if he knew that his feelings were still that strong for her, then he knew that he needed to change his mind.

A couple of days later, one of Kyrie's cousins, Valerie, visited Christiana. She told Christiana that Kyrie's fiancé had been calling him for the past two weeks and that she could not catch him. Valerie also said that she knew that Kyrie was making a mistake by going through with the marriage because he was still in love with her.

Christiana told Valerie that she knew that he was still in love with her, but he would probably go on through with it because he wouldn't have the courage to cancel. Christiana also said it was too close to the due date and that she was pretty certain that his fiancé's tearful plea of 'how could he do that to her' and 'with all the money that has been spent' would seal their deal.

Kyrie's fiancé also found out that he had been spending time with Christiana because she was informed by one of his sisters when she called to question his whereabouts. The sister informed her that the reason she was not able to reach him was because he had been spending time with his ex-girlfriend. Eventually, he was questioned by his fiancé, and he told her whatever he needed to tell her to cover up his actions. Ironically, Kyrie became upset with Christiana instead, but it was only because he had gotten caught.

Christiana was the bigger person with him because she knew the truth. Christiana went and delivered a congratulations card for his upcoming nuptials; however, Kyrie acted agitated by her friendly gesture. Once Christiana took notice of his unpleasant demeanor, she also understood what he was upset about, and that was because she was the one who had his heart.

Christiana made it easy for Kyrie to walk away from her by not questioning why he was going through with the nuptials. She was not going to add stress nor drag him, knowing that her actions with him were pure. Christiana also knew that he would regret not making the right choice and that he would be living a lie.

Christiana was in for another surprise. A month after Kyrie got married, he arrived back in the city and once again made his way to look for her. Once he arrived at Christiana's home, he rang the doorbell. She went to the door to answer it and saw him. She did not bother to step out, nor did she offer him to come in because she did not like the fact that he was that comfortable showing up after how he acted towards her.

Kyrie said, "I just wanted to stop by to say hello, and I wanted you to meet a few of my military brothers." His military brothers were in the car, waiting and watching to get a glimpse of her. Christiana looked out at the car briefly as Kyrie saw that she was not up for the nonsense of him showing her off to his friends. She thought to herself, 'What the hell is he doing?' Christiana guessed that he wanted his military brothers to see what he had let get away. After that visit, every time Kyrie made a return visit home, he managed to make a beeline straight to find and see Christiana.

Chapter 29: Dream Crushers

Christiana was ready to go off to college until Nadia, who had no interest in going to college for an entire year, realized that Christiana had made her arrangements and was ready to go. Bash called his daughter and sat down to talk to Christiana. She was informed that he could not afford to send both of his daughters to college, and because Nadia had graduated first, she should be the one to go first.

Christiana could not believe that her dream of attending college was over in less than a five-minute notice. Christiana had to toss herself into the workforce. The first thing she tried to get was a modeling career because she had done a couple of modeling talent shows, which went well for her.

Nadia immediately went off to college, but she returned by her second semester pregnant. Bash then talked to her baby's father, Damon, and made him an offer that he would not be able to refuse. Bash told him that if he were to marry Nadia, he would give him a job at the post office because he had that sort of influence while making sure they were going to smooth things over for Nadia at all costs, but that would eventually set her up for failure.

Chapter 30: No Holds Barred

One evening, Christiana arrived home before her father arrived home from work, and little did she know it would soon end in an argument that would cause her to move out. Natalie came in and started an argument with Christiana over the fact that she had prepared a meal for her father and herself. Natalie had put on some hotdogs with pork and beans in a pot before she went on a fishing creek, or at least that was what she had told her husband that morning before he left for work.

Natalie was actually going out with Scott. She then threw anything into a pot for when her husband was to arrive back in from work to eat. Once Bash arrived back home, he looked inside the pot and said, "I don't want this shit." Christiana said, "Yeah, me neither!"

Christiana then reached up into the freezer and pulled some steaks out. She then made some other sides for them to enjoy with their meal. Once Natalie arrived back home late and to avoid being questioned, she decided to start with Christiana. She said to her, "If you don't like the shit that I fixed, then fix the shit yourselves!"

Christiana looked confused at her mother because they were already eating the meal that she had just prepared, but she then said, "That's what I did!" Bash chuckled at his daughter's response, but he also thought, oh shit! It's about to go down! Bash's chuckling seemed to have made Natalie even more mad with Christiana.

Meanwhile, Nadia and one of her friends were also present at that time. Natalie decided to go in with a low blow for their entertainment purposes. Natalie said to Christiana, "Yeah, with your bull-daggering ass!"

Christiana looked at her mother disgustingly and said, "That's a horrible thing to even say to your own child, but seeing how you really feel about me, I will be out of your house come this Friday!" Christiana realized that it was something that her mother and Scott had to have conversed about before because he was forever talking about bull-daggers.

Once Friday arrived and after Christiana left work at the end of her shift, she then pulled up with boxes ready to pack and go. Christiana was able to get one of her close girlfriends to agree to move into an apartment with her and share the cost with her.

Shortly after the move, Christiana was at home one day waiting for one of her friends to come to pick her up so that they could go to lunch together. Christiana heard a knock at her front door. She went to see who it was, and to her surprise, there stood Natalie with her aunt Adrian and her uncle Raymond, her mother's brother, with Scott tagging along behind with some cold beers in hand.

Christiana immediately thought to herself, "Oh hell no, this won't be one of your hanging meet-up spots!" Christiana told her mother that she was about to leave because she had plans. In other words, to let her know that she wouldn't be using her. Natalie said, "Oh, we were just stopping by," but her mother also saw that it was not a warm fuzzy welcome by Christiana's facial expressions.

Christiana's Uncle Raymond saw the look on his niece's face and said, "Don't worry. We are about to head over to my house, so I guess we'll see you later!" Christiana said she would stop by her Uncle's house after she was done having lunch with her friends. Once she walked into the house, Christiana saw everyone but her mother and Scott. That was when she asked everyone where her mother was.

Adrian pointed towards the basement, then angrily said, "She down there!" Christiana knew from her aunt's comment that something was not good, so she went down the stairs to see what her mother was doing. When she got down there, Christiana saw her mother crying big crocodile tears as if she was pleading with Scott. This sickened her and made Christiana mad at her mother. She didn't even bother to ask what was wrong because she knew her mother was already in the wrong by being with another man.

Christiana immediately went back upstairs and then told everyone present that she was out of there while shaking her head, letting them know how pathetic it was. Later that evening, when Christiana arrived back at her apartment, her roommate said, "Your father has been calling here looking for your mother. As a matter of fact, he's called a few times now." As soon as she said that, Christiana heard a knock at the door.

Chapter 31: Loss of Respect

Christiana went to the door to answer it, and when she opened the door, there stood her mother looking like a night of sex and regrets. Christiana said to her mother, "Your husband is looking for you, so you need to bring your ass on because I'm taking you home right now!" Christiana let her mother know that this was the first and the last time that she was going to get her involved in her shit, and she meant what she was saying.

On their way, driving her mother home, Christiana kept lighting into Natalie's behind. Once they pulled up in the driveway, she told her mother, "Now you are on your own, and I'm not coming in there to lie for you either! Just know that if Daddy calls to ask me anything, believe me, I'm telling him that I don't know where you were!"

Chapter 32: Lights Camera Action

On her ride back to her apartment, Christiana pulled up to a stop light. While sitting and waiting for the light to change, another vehicle pulled up beside her. In that car was an attractive guy, and they both looked over at each other. Each one smiled and then waved. Christiana figured that she would switch lanes and get behind his car as they drove. She figured that if he wanted to meet her, he would eventually signal for her to pull over.

To Christiana's surprise, he did just as she had suspected, not realizing that it would be a life-changing mistake that she was about to make. The guy slowed down and then put his blinker on, showing that he was going to pull over. This guy was letting Christiana know that he wanted to talk to her; she then followed suit and pulled over about a car length behind him.

The moment the guy exited his vehicle and headed back to Christiana's vehicle, things went into slow motion, and what she saw did not make sense to her. It was like she was looking at live movie clips.

The Most High showed Christiana everything that she needed to see and understand about the guy. Christiana was shown that this guy was a wicked, evil player; she also saw that he had a family where he was not good to the woman or the children. Christiana also saw that he loved clubbing and that he was a womanizer, but what she did not see was that it was going to be her.

Christiana, at that time, was not spiritually awakened, and therefore, she did not know that the Most High was warning her. Once he made his way to her car, Christiana rolled down her window so that they could officially meet. That was when he introduced himself to her, stating that his name was Di'Ban. Christiana first thought to herself that it was weird, but she then went on and introduced herself to him.

Di'Ban said, "I was hoping that you would have pulled over so that I could talk to you." He then asked her if she was single. Christiana responded by asking him, "Are you single?" Di'Ban told her that he was at that point. She told him that she was, too. Di'Ban asked if he could get her phone number and that he would call her later that evening.

Christiana decided that she would give him her number now. As she was jotting her number down, she double-checked it to make sure that it was correct before she handed it over to him. Unbelievably, she had given him the number incorrectly. Christiana went home to her apartment and told her friend that she had met a guy on her way home. She then went on to tell her roommate what she had seen before they spoke to each other. That was when her roommate said, "Girl, make sure that you don't smoke any more of that shit because you are tripping."

Christiana said, "No, seriously, it was like a movie was playing in slow-motion clips!" Christiana got annoyed trying to explain it. She then said, "Anyways, he said that he was going to call in a little while, so I will see how it goes." Christiana was kind of hoping that he didn't call because what she saw bothered her.

Christiana did not receive a call from Di'Ban that evening, but the first thing early the next morning, Di'Ban came ringing her phone, waking Christiana up from her sleep. He asked if it was Christiana. Once she stated that it was her, he said, "This is Di'Ban." He also said, "Listen, I told my roommate, I know this woman did not mean to give me the wrong number, and I literally have spent all night turning these numbers around until I just reached you."

Christiana thought to herself, how strange. Once he stated that he had spent all night turning numbers inside out to reach her, Christiana apologized, stating that she didn't know how that happened because she knew she had double-checked the numbers.

Di'Ban asked Christiana if she had any plans for the day. Again, she thought this dude seemed anxious and that his behind should be tired if he was up all night. Once more, she did not listen to the warning signs.

Christiana said, "I was planning on going to this linen store to buy some items for my bedroom because I just recently moved into my apartment." Di'Ban immediately asked Christiana if she would mind if he joined her. Once more, she was thinking, 'Damn, he must be one lonely guy or just eager to see me,' but finally, she said, "Yeah, ok, that's fine!"

Christiana told him that she was planning on leaving around 2 PM that afternoon. She let him know that he could meet her there at the store if he wanted, but he said that he could pick her up because that was not a problem for him. Christiana agreed to let him pick her up and then provided him with her address.

Christiana expected him to arrive at her home somewhere around 2 PM to pick her up.

Chapter 33: Red Butterflies

Christiana loved eating all sorts of foods and never had any food issues that bothered her stomach. She had prepared herself a meal, and while enjoying her meal, she heard a knock at the front door around 1 PM. Christiana's roommate told her that she could keep eating and she would get the door. Once the front door opened, there stood Di'Ban.

Once Christiana looked up and saw him, her stomach immediately began flip-flopping. Christiana's roommate noticed that she couldn't finish her meal, and she had started putting her plate away. Meanwhile, she looked at her roommate and then said, "Strange, huh, you know that I can eat regardless.

After the shopping trip was over and they were heading back to her apartment, Di'Ban asked Christiana how long she had been single. She said, "Long enough," but she was not in any rush either. Di'Ban stated that he had been in a long-term relationship and that it was over. She listened to every word, but she also thought, let's just see where this was about to go.

Di'Ban mentioned that he needed to stop by his apartment. After they were headed back and arrived at his home, he asked Christiana if she would like to come in. Christiana accepted the invitation to go in, but all he wanted was for his roommate to see how beautiful she was.

After that visit, Di'Ban immediately made it a point to call and visit Christiana damn near every day. They had not become sexually involved yet, but he always managed to spend the night by staying late until they both fell asleep.

Christiana came in from work one day and noticed that he had one or two items that he had left behind at her home. She immediately took his items and neatly placed them on the couch near the front door for him to see upon his next visit. Christiana wanted him to know and clearly understand that he was not going to try to pull a sneak move in on her.

Shortly after packing up all his attire near the front door, Christiana's roommate came in from work. Once she entered the door, she immediately burst out laughing loudly and said, "I see, you do not play!" Christiana responded loudly and said, "Nope," without even asking what her roommate meant by her comment. Christiana went on to say, "I think his slick butt was trying to be sly with it, but no, ma'am, he needs to take his belongings to his own domain."

Di'Ban showed up later that evening and saw his belongings near the front door. He immediately mentioned that he was intending to get his things. Christiana lightheartedly said, "Yeah, I figured I would bring them up here for you so you don't forget because it was looking like you were living here."

One day, Di'Ban asked Christiana if she would like to go to the gym with him to watch him and some of his guys play basketball. Once she stated that she would go, he made it a point to take her with him each time thereafter as if she were some sort of a showpiece.

Christiana had a girls' night out scheduled at her apartment one evening. Everyone had already arrived on time other than Christiana because Di'Ban had started consuming a lot of her time, which was why she was late for her own affair.

Once Christiana arrived, everyone got a chance to meet him because he walked Christiana in before he finally made his exit. After he left, most of the women were saying he seemed nice and charming, but that was when Christiana sarcastically said, "Yeah! Yeah! Yeah! Listen, there's something about him that seems to just creep me out!" All the ladies surprisingly made a gesture by saying, "What?" Christiana said, "OK, the next time you all see him, just take a good look at his eyes. He looks like he could be clinically crazy!"

That comment sent everyone through the roof, and a couple of ladies even went as far as to say, "Girl, you are crazy, but if you don't want him, then I'll take him!" Christiana was going to learn that she should have taken their offers that day.

Christiana's birthday was soon approaching, and Di'Ban told Christiana that he wanted to take her out for her birthday to hear a live band play. Christiana agreed that it sounded like a nice outing for her birthday, but what she didn't know was that it also involved him bringing his mother along on that date. Christiana had never been on a date where someone actually brought their mother along.

Di'Ban picked up Christiana first and then went to pick up his mother. Di'Ban and his mother were making their way to the car, and one could see by the look on his mother's face that she was hoping to see if

Christiana was going to get to the backseat. Right before they made it to the car door, Christiana had already opened the door and jumped in the rear seat because she had respect for anyone's parents.

Di'Ban and his mother entered the car, and he immediately introduced Christiana, but his mother had a rough edge when it came to engaging in any pleasantries. Christiana said, "It's a pleasure to meet you." However, she was unable to catch his mother's name because she had turned up the radio to keep Christiana from hearing her and her son's conversation.

Di'Ban pulled up to the club to let both ladies out at the door, and that was when his mother began taking an inventory of everything that Christiana was wearing from head to toe. Christiana was asked about the fur coat that she was wearing and as to where she got it from, but she played it down, stating that she had it for a long time and that she really didn't remember, hoping to keep the attention low on herself.

Di'Ban walked inside the club with both of them, but then there came another family member, who was also not very warm or welcoming, to meet Christiana. Once they all had made their way to a club table, Di'Ban asked Christiana if she would like a drink, and at that point of uncomfortableness, she was ready for a drink.

Christiana saw that all eyes were on her and that it was going to be a rough evening. However, she had planned on making the best of the evening. The waitress came, and Christiana ordered her a stiff drink on the rocks. His mother was paying close attention to what and how much

Christiana was drinking. She was also keeping count of how much money her son was spending on Christiana.

Christiana could tell it was about to be interrogation time between his mother and sisters, but Christiana knew exactly who she needed to call to level the playing field. She called her parents, who also brought one of Christiana's cousins. Once they arrived at the club and took a seat, Bash had a way of making small talk that seemed to calm those at the table. Once the club was about to close, Bash invited them all to come to their house for breakfast.

Once they arrived at the house and while Di'Ban's mother took a stroll, looking through their home, she and Christiana had a meetup in the family room. That was when his mother asked Christiana, "So, how old are you on this birthday?" Christiana stated that she was 21, and his mother said out loud to her son, "Di'Ban, What are you doing with this young broad?"

Christiana was shocked that a woman of her age could be so ignorantly rude. Christiana was certain that her mother had to have heard his mother's comment. She also knew that her mother felt it was okay if she had said something disrespectful to her daughter, but not for some other woman to be disrespectful, especially in her home.

Once everyone had eaten breakfast, they all sat in the living room and talked. Di'Ban's mother sat on one of the couches that faced Bash. This woman pretended that she had fallen asleep and had her legs slightly cocked open with a dress on so that Bash would be able to see between her legs. Christiana kept noticing that her cousin was staring at her and then back at his mother. After the third attempt to get her to look, Christiana finally noticed, but by that time, so had Di'Ban. Her son knew that she was not asleep but was trying to get Bash's attention. Di'Ban then called out to his mother, telling her to let's get ready to go.

Chapter 34: Living for Others

Christiana had taken on a new modeling job, but Natalie and the rest of the family seemed to have had a problem with her assignment as it involved modeling lingerie. Natalie made it a point to drag Di'Ban into their dislike regarding Christiana's job. However, what they did not know was that she had been scouted out by a known modeling agency in the city. Truthfully, they feared that Christiana might make something more of herself.

Di'Ban felt that because he and Christiana had been seeing one another exclusively for a couple of months. He also let her know that he did not want his girlfriend to be seen in lingerie. Christiana did not recognize that she was trying to please people and did not think of adding herself to the equation for what was best for her.

After about a year of dating, Christiana believed the relationship was going great. Until she found out that she was pregnant, Christiana had other plans as far as career options that she wanted to embark upon. But again, he had an issue with that because it would require a lot of travel for her.

Once she told Di'Ban that they were about to have a baby, he took on an entirely new character, and it changed the relationship. Christiana already knew that he had a daughter, which he had denied, stating that he was not sure if the child was even his. However, once Christiana saw the picture of the child, she said, "You do know that there is a DNA test that can be done and that will provide you with the answer, but as far as I can see, she looks exactly like you."

Christiana asked, "What are you going to say when she shows up at your door one day? How will you explain if I answer the door standing there with a child? How will you justify that towards your daughter?" However, he had no response to her questions. Christiana was about three months pregnant and was dealing with a horrible case of morning sickness on top of Di'Ban's nasty attitude about the pregnancy. He didn't care about Christiana herself dealing with how her life plans were about to change.

Christiana was in the house with Di'Ban one evening when the doorbell rang. He went outside but did not come back in anytime soon. That was when she went and looked outside and found him sitting in the car with another female, appearing unphased and unworried.

Christiana picked up her things to leave, and not once did he get out of the car to inquire as to where she was going, nor did he attempt to go and look for her once he realized she was not returning.

That same night, Christiana went and stayed with a friend, who suggested that she might be overreacting. So, she went back the next afternoon. Once she had gotten into the bed to take a nap and later woke back up, she then pushed the sheets back and saw what looked like blood on the sheets. Christiana jumped up and checked herself thoroughly and found no sign of blood coming from her. She then noticed that it appeared the sheet had the look of being wedged between someone's butt cheeks. Once she brought it to Di'Ban's attention, he told her that it had to have come from her. Later that same afternoon, Christiana found a used paper napkin in the bathroom trash that had a woman's lipstick marks on it that did not belong to her either.

Chapter 35: The Skeletons Closet

Nadia called Christiana and said, "I need to talk to you about something." Christiana said, "OK, I'm listening." Nadia said, "I have talked to our mother about her relationship with Scott. I am tired of it and having him sitting with his feet up under our father's table eating daily." Christiana asked, "OK, so what did she say? Because I, too, have been sick of the shit a long time ago."

Nadia said, "Well, I think that we need to go and have a talk with Daddy in private. I know that if we ask him to keep it private, he will, and I don't think that he would do anything absurd."

Christiana asked, "Well, when did you want to talk to him?" Nadia immediately three-way called their father and then told him that she and Christiana wanted to have lunch with him the following day, and their father said that it would be fine.

Nadia picked up Christiana for lunch, and when they arrived at his office, they saw that Bash was glad to see them. Once they went in, Nadia directed Christiana to close the door behind her. Nadia said, "Daddy, we didn't really come for lunch; we really came just to talk to you about something." Christiana was expecting Nadia to continue talking and get the conversation started because she was the one who called and suggested talking with their father. Nadia immediately stopped talking and then looked for Christiana to spearhead the rest of the conversation.

Christiana was as tactful as she knew how to be, but she was not the one to spin her wheels or sugarcoat matters when it came to serious conversations. Christiana saw that Nadia was going to leave it all upon her to break open the conversation.

Christiana said, "Daddy, what we really came here for is that Nadia and I first need you to make us a promise that you will not go home and do anything that would land you in jail. Can you make us that promise?" Bash looked confused and humbled but then agreed that he would promise them, which set the tone for the conversation. That was when Christiana said, "Nadia said that she had already talked with our mother, but she felt that we needed to talk to you because of the nature of the issue." Christiana looked for Nadia to jump in to help with the conversation. However, Nadia left her sister hanging. Christiana realized that it was too late to backpedal out of the conversation. She said that they were there to let him know that they were tired of the disrespect that had been going on in the home, which had been coming from his so-called friend Scott and their mother. Christiana said, "Daddy, they have been involved in an affair with each other, which has been going on for way too long, and we are sick of it!"

Before Christiana could say anything else, Bash said, "Well, I have something to tell you girls too. One of you girls is not mine, and Christiana, I know that you think that it is you, but no, Nadia, it is you, but I have loved you like you were my daughter."

Christiana sat there and had a high-in-sight moment thinking, what would make him say, "I know you think that it is you, Christiana." That comment confirmed what Christiana had always felt, that she had been treated indifferently, but she came to understand the reasons, which were even more hurtful. Bash then added insult to injury and said, "Christiana, we were not happy about your pregnancy either." Christiana realized, in that very moment, that her father had ostracized his own daughter while he uplifted some other man's seed to cover up Natalie's shame. That comment set the stage for his daughter to stay in a dysfunctional relationship that she should have run far away from. Christiana stayed in the relationship with the hopes that her parents would eventually approve of her family.

Bash immediately left work and went home to talk to Natalie. Once he got there, he told her that both the girls had come to him and informed him that she and Scott had been involved in a long-time affair. He told her that she would have to end the affair now, or he was going to divorce her because he had had enough of her cheating.

Natalie began to lie. She told Bash it was not true, that it was Christiana who had made up that lie, and that Nadia had never come to her talking about any such thing. She also said, "I do not want Christiana to come back over here either with her lying ass."

Natalie then jumped on the phone, mad as hell, and called Christiana while Bash stood on the sideline. She said, "You little lying ass bitch! I don't know why you would tell your father something like that because you could have gotten someone killed." Natalie went on to say, "I don't ever want you to bring your ass back over to this house," then hung up the phone. Christiana listened closely and realized that her mother never mentioned Nadia while she solely raged her wrath on her. Christiana called her sister after that phone call and told her what their mother said to her. Christiana asked Nadia if their mother had also called her.

Nadia said, "No, she did not call me." She didn't attempt to care that Christiana was the only one who got blamed for what Nadia had orchestrated. She was fine with letting her sister be the one to be the fall guy all by herself. Even Bash did not attempt to clear up the fact that Nadia was the leader in revealing the affair that she drew Christiana into.

Bash saw that Christiana wasn't coming around anymore. He called her one day to tell her that she could come to the house to visit because he did not care what Natalie said to her, stating that it was his house and she was welcome to come whenever. Bash had a BBQ one weekend. He called Christiana to tell her that he expected her to come forcing her decision to come back. She did go to attend the BBQ, but she could feel the tension between her and her mother, who hardly said anything to her the entire time.

The months passed fast for Christiana. She was in her 8th month of pregnancy. She decided to take part in Lamaze class to prepare herself for childbirth. Christiana could not get Di'Ban to go with her because he kept making all sorts of excuses for not attending the class. However, there was one of Christiana's cousins who agreed to fill in and went with Christiana each week.

Christiana was at her parents' house visiting one day when Kyrie came back to town, and then he made his way to see her again. To Kyrie's surprise, he saw that Christiana was with child and that they wouldn't be able to partake in their usual walk together. They both kept the conversation very light, not talking about anything other than what he had been doing. Kyrie looked shocked to see that she was expecting. He left shortly after arriving, but he was still not done with Christiana.

Chapter 36: Spin on Your Heels and Run

Christiana was much like her father in trying to keep a family together. She was putting up with a lot that came with big sacrifices. Christiana was learning all about Di'Ban's true character, and she found out that he was one unnecessary drama-filled, uncaring, inconsiderate, uncompassionate, selfish, wicked, did not give a fuck, woman beater, non-legacy leaving, forcing himself upon her, questionable, stalking, lying, and the truth was not in him, using the Most High's name for his gain, conniving, moral-less, should have been snipped at the birth, piece of shit, ass of a male.

Christiana ended up asking her friends to help her with painting and putting the baby's room together before the arrival of the baby. Christiana's debut arrival was finally at hand for the delivery, and just before wheeling her into the delivery room, Di'Ban suddenly decided he wanted to go in the delivery room. Christiana said no to him because her cousin was faithfully there with her through those Lamaze classes, helping get her through that labor. She was not going to rob her of that moment that she had worked for.

The day that mom and baby were released to go home, and the family had come to see the baby, Di'Ban took advantage of them being there and left for a basketball game, leaving Christiana to fend for herself with their newborn son. Christiana named the baby Cassius because he came into the world fighting through twenty hours of long labor.

One thing that Di'Ban learned about Christiana was that her father had raised his daughters and taught them to be self-sufficient for any case of survival. Christiana had paid very close attention to her father's teachings, and it was what she knew that got her through, even when she was standing alone. Christiana recognized that she was pouring into an empty vessel and that she was no longer interested in that relationship. However, Di'Ban, who thought that he was irreplaceable based on the fact that she had his child, was going to eventually find out that once she was done, she was done.

One day, they were out taking a ride, and Di'Ban began to start an argument with Christiana. He made the mistake of telling her that the relationship was not shit. Christiana looked at him and said, "You know what, you are absolutely right." Christiana told him, "I will be out of your house in a couple of weeks." To Di'Ban's surprise, at the end of that second week, one of Christiana's friends came from Atlanta with her truck and they began loading up to move Christiana with the baby to her new apartment. Di'Ban realized at that moment that Christiana was not a bluffer, and she spoke literally. He immediately decided that he would help with the move, but it was only to see where she was moving to.

Once they had made their way into the new apartment, Christiana caught him staring out of the patio door in a disbelieving daze. Each day thereafter, Di'Ban kept making his way back over to her new place. It was to the point that Christiana thought, what the hell did I even move for because he won't stop coming over here? Christiana knew she was done, but she did not know how to tell him that she was no longer interested in him or any relationship with him.

Somewhere around the third or fourth month after the move, Christiana was feeling down and depressed with the entire situation. One afternoon, they were riding in the car, and Christiana was in a deep thought while staring out of the window in a daze. However, Di'Ban must have felt her energy because that was when he asked her, "What are you thinking about over there?" That was when it snapped her back to reality. She looked over at him and said, "Nothing," and realized that he even wanted to know her waking thoughts, even for a man who felt their relationship was not shit.

Di'Ban had gotten comfortable with the living changes and was back on the prowl with other women. One evening, Christiana found out about it when she went for a visit. Di'Ban had gone to pick up a pizza for them, and that was when another female called his house. Christiana answered the phone, and when she asked who this female was, she told Christiana that she and Di'Ban had been seeing one another for about two months.

Christiana confronted Di'Ban about it and told him that she was done. He lied as usual, but once he saw that Christiana was moving on without him and was not giving a damn about what or who he was seeing, his ego could not take it.

Di'Ban showed up at Christiana's place late one evening, apparently to see if someone else might have been there because he had not called her to check if she wanted to see him. Once he knocked on the door, Christiana looked out to see who it was. She became annoyed that it was him. She said, "This son of a bitch is a damn stalker!"

Di'Ban was allowed entry. He then started claiming that he wanted to see his son Cassius. Di'Ban went and woke his son up as Christiana remained in the living room, hoping that it was going to be a short visit. He then put his son right back to bed. Then, she pleaded with Christiana, talking about how he missed both her and his son and that he was ready to be a family man. Di'Ban could see that Christiana was not interested in anything that he was saying, and once all else failed with pleading, he resorted to crying. Christiana knew deep down inside that she did not want to give in, nor did she want to be with him, but against her best judgment, she gave in to his fake antics of tears.

Di'Ban went as far as taking Christiana to shop for an engagement ring, and when Valentine's Day arrived, he placed the ring into a box of candy. Unfortunately, Christiana fell for the okie-dokie bullshit.

Christiana immediately started putting a wedding together. She even found her dress. Now, just six weeks before the wedding was to take place, Christiana, due to carelessness on her part, became pregnant again.

Christiana thought that it was no big deal because they were about to get married. However, when she informed him that they were expecting another child, he gave Christiana an ultimatum of aborting the pregnancy, and if she chose not to abort, then he would call off the wedding. Christiana right then pulled the ring off her finger and placed it on the table for him to take with him.

Once the ninth month arrived, Christiana had to call and plead with Di'Ban to come over one night because she had become very irritable and needed help with Cassius. Di'Ban did not make it easy but he did eventually arrive.

Christiana then took to her son's bedroom, and just as she was starting to fall asleep, the first labor pain hit her. At first, Christiana thought it was from something she had eaten earlier that day. However, once she went into the bathroom, another pain hit that literally took her breath. Christiana called for Di'Ban more than once as he took his own sweet time getting there.

When Di'Ban finally opened the door, looking annoyed as hell with her, she said, "I am in labor!" Christiana asked, "Can you please call the OB Doctor?" But Di'Ban played a stupid role even at a time of urgency. Christiana said, "Just give me the phone, damn!" Once she reached the doctor, he told Christiana that she needed to get to the hospital right away because her contractions were about a minute and a half apart.

Christiana called her sister Nadia to see if she could come to help with Cassius. Nadia made her way to her apartment while Christiana tried to make her way down the stairs between contractions without any assistance from Di'Ban.

Once Christiana made it into the OB area, the OB Doctor said that he could see the baby's head and all he needed to do was to bust the water bag. The OB doctor then told Christiana that once he burst the bag, all she needed to do was push. The OB Doctor called for Di'Ban as he sat in the waiting area along with Nadia and his son. Di'Ban then acted as if he did not hear the hospital staff calling for him.

Nadia later told Christiana that she had to tell him that they were calling for him. She also said that Di'Ban foolishly said that his name was not Mr. Bouvier before he made an exit.

Once Christiana's water bag had burst while she was on the birthing table, the Doctor told Di'Ban, "You will need to help her!" The Doctor said, "Listen, once her next contractions start, I'll need you to raise her up so that she can push each time through the contractions, which you can tell by watching this machine to see when they are starting!" The nurse told him, "Please make sure to keep watch to make sure she keeps breathing because she is having this baby naturally; she is in excruciating pain, and there is no time for us to give her anything to ease the pain of this labor."

As soon as a contraction would start, Di'Ban would raise Christiana up and then carelessly just let her drop back down onto the birthing table. Christiana couldn't even try to say anything to him about his lack of empathy before another contraction would hit.

Christiana's pain from the contractions was so intense that she did not even feel the Doctor once he needed to cut her to receive an episiotomy. Meanwhile, she was still begging for medications from the contraction pain. Christiana did not recognize that she was already well past that part of the labor stage.

Christiana was in what was known as spontaneous labor and delivery. She was still fully dressed when she delivered a second baby boy 45 minutes later from the start of that first contraction at home. Christiana named him Niles because the OB doctor kept trying to keep Christiana focused during her delivery by asking her what she was going to name her son. Christiana asked the doctor, "What is your name?" and he said, "Niles." Christiana said, "His name is Niles."

After the birth of the second child, Christiana quickly gained back her beautiful luster, and she returned to work shortly after giving birth. She and both children were doing great. Christiana went and gained child support because she found out that had she not gone to set it up. Di'Ban would have left every bit of responsibility upon her as he had done with his daughter.

Niles was only a few months old when Di'Ban called Christiana, asking if she could bring the children over to his house, where he was having a family BBQ. He wanted his family to see Cassius and the new baby because they all had not seen Niles before. Christiana took the children for the visit, and once she arrived and then walked out to where everyone was, she was greeted by chilly welcomes and stares.

Di'Ban's brother James walked over to Christiana after seeing how everyone was acting and said, "Come on, let's go inside." James said, "These fuckers are sick. They just wanted to see if Niles looks anything like his behind." James also said, "I heard him tell them already that he didn't think that Niles was his." He quickly grabbed up his nephews to bring them inside with Christiana.

Christiana first chuckled at the foolishness and then said, "Well, how interesting is it that he looks just like his grandmother." Christiana didn't stay long for that visit because she did not desire to sit amongst shitty-ass, acting people, nor was she a fan of faking pleasantries.

Chapter 37: Cheaper to Keep Her

Di'Ban figured that he would try to get back with Christiana regarding what he thought was a crafty plan. He didn't understand that her concern was for her children's overall well-being. He would show up, taking her and the kids to certain events like fireworks shows or toy stores. However, he realized Christiana valued family unity, and he was playing that role much like a chameleon who could change on a dime to fit any occasion.

Di'Ban believed that he was smarter than Christiana, but what he didn't know was that he was showing her along with teaching her what game playing was all about. However, eventually, he was going to learn that he would play himself in the end.

Once again, he came asking Christiana if she could move back to his house so that he could be with his family. Di'Ban always had ulterior motives, but this time, it was to see if he could get Christiana to drop the child support.

The plan he conjured up was not going to work because Christiana would not stop the child support because she knew what selfish game he was playing. Soon after the plan unfolded, he was out running her down to people while playing the victim. This time, he told everyone that he could not believe that she still had him to pay child support for the children while living with him.

Di'ban was forever negatively running Christiana down to his friends until, one day, his friend Diego came to visit him. Di'Ban was bragging to Diego that he was going to visit Atlantic City for a week and was not taking Christiana, but he was taking another female.

Once that weekend arrived and he left for his trip, Christiana was at home with the kids relaxing when the doorbell rang. It was Diego, Di'Ban's buddy, who knew that he was out of town. Christiana opened the door, and Diego stepped in, stating that he was checking on her and the kids. Then, after a few minutes, Diego said, "You know, Di'Ban is a fool with a woman as fine as you, and I know you're good to him. I just don't know how he could leave you here." Christiana right then decided to put her children to bed but before returning to the living room, she made a quick pit stop into the kitchen. Christiana picked up a container of lighter fluid and an old-school zippo lighter before she returned to the living room with Diego.

Once Christiana returned, she stood at the doorway, and that was when Diego said, "You know if you were my woman…" but before he was able to finish, Christiana said, "Yeah, well, I appreciate your compliment and all, but you knew that Di'Ban was gone, right?" Diego was shocked by her rapid response.

When Christiana pulled the lighter fluid and zippo from behind her back, she harshly popped open the fluid cap and then flipped the lighter top open while looking directly at him.

Christiana sarcastically asked, "Diego, did you really think that I would fall for that bullshit?" Diego saw the items in both her hands, and she saw the fear in his eyes, hoping that she wouldn't douse him with the fluid and set him ablaze. Christiana went on to say, "Now, what you need to do is casually walk your slithery behind out of here before I cause your ego and your beauty to fade."

Once Di'Ban returned from his trip, Christiana said, "I hope you had a great time; however, while you were out with whoever in the dafux you were with, oh yeah, and while you were running your mouth to your friends, I know you didn't realize that you had left the door open for Diego to comfortably think that he could come to your home to hit on me. Christiana made sure to say, "Now, you really can't be mad at nobody but yourself, but trust me when I say I already handled your sloppy-ass mishandlings by myself!"

Di'Ban tried to angrily say, "I just know that motherfucker didn't think he could come show up at my house…" but right before he could say anything else, Christiana said, "That's what happens when you run your mouth, leaving your bed open to slide your ass in another woman's bed so you can't be mad when you flock with the same kind of birds!"

Christiana had made a foolish decision to remain in a loveless relationship for the sake of the children. She was doing exactly what she had seen her father do with her first-blood family. Christiana respected her father's sacrifices, but Di'Ban's sick energy had begun to take a heavy toll on her.

The boys were growing very quickly. Cassius was five years old, and Niles was three years old. Christiana realized how much of her youthful time she had already wasted with an unhealthy relationship.

Christiana had lost her expectations of Di'Ban, considering what he had already shown her. Christiana was constantly reminded of what she had been shown from the moment when she first saw him.

Christiana had secured a new place for her and the kids, but as soon as she had, Di'Ban announced some devastating news. He found out that his mother had fallen ill and that her days were numbered. Christiana, being a compassionate person, convinced herself that she should not leave him at a time like that because it was his mother, and he was about to experience a loss. Christiana stayed even after she had secured her new place, which Di'Ban was unaware of.

Christiana had a close neighborhood friend, Sicarri, whom she had known her entire life. He returned to the city along with his girlfriend Angie after leaving the military. Sicarri and Angie moved back to Chicago, but they needed a place to stay. Christiana, being in a position where she could help, offered to let them stay in the apartment until she was ready to make her move.

Christiana stayed supportive of Di'Ban when it came to his mother's critical health diagnosis, regardless of how his mother treated her. One day, Christiana got his mother a beautiful bouquet of roses from her grandchildren as a peace offering to show that she cared, but his mother didn't even bother to open them. Christiana understood that no matter what she had done, his mother would rather die being as mean as a rattlesnake towards her, and nor did she ever take to her grandchildren.

Christiana remembered washing his mother's shit-filled drawls out because she was not able to control her bowels, and not once did his mother say thank you. Christiana also realized that Di'Ban did not appreciate that she was giving unselfishly to support him at a difficult time of their lives. He still ran her down, but she understood that she was cut from a different cloth.

Di'Ban told everyone that Christiana was helping him so that he would marry her, but he was too foolish to realize that she only made the sacrifice for his period of grief. Christiana was about to depart from his life soon because she had heard all the lies that he was spreading against her. Meanwhile, he was trying to make himself look like she could not do any better than him, but he again was about to be surprised that she was on her way out and how well she was about to be living.

Di'Ban's mother died a few weeks later after her diagnosis, and when the wake came, Christiana was amazed that even in death, how vain Di'Ban and some of his family were, making sure to dress their mother up in a bunch of jewels, but not understanding that those jewels did not change the person that she was. There was no substance of character that was even spoken about his mother during her services.

Christiana had never thought about where a person's soul would go upon death until one of her relatives, Julius, who enjoyed being around Christiana, approached her. While out on a walk, he asked exactly what Christiana had been thinking. Julius asked, "Where do you think her soul went?" while looking wide-eyed and chuckling because he knew that Christiana had already thought the same thing as him.

Christiana shockingly said, "Oh shit, I thought I was the only one who had that thought." All she said was, "Listen, I have no idea!" I was just hoping that she had made peace with the people with whom she knew she had been a ferocious bitty towards before she drew her last breath.

Shortly after the death of Di'Ban's mother, he was back to his nasty ways. However, that was when Christiana let him know that she was about to move out. Di'Ban could never understand how she always seemed to manage to land on her feet, not knowing that she had grown up learning how to survive since she was a young girl. There was one thing that Christiana knew well, which was how to plan and pack to make things happen for herself.

Later in that year, Cassius started kindergarten, which was only a half day., Christiana would take and drop him off at school on her way to work. This schedule would not allow for her to drive her son from one side of town to another to drop him off for a sitter and then for her to arrive back to work on time.

Christiana asked her mother, Natalie, that if she paid her, could she pick Cassius up at noon from school and then take him home with her until she got off from work? Natalie told Christiana that she could not do it without excuse and left it up to Christiana to figure out.

Christiana could not get anyone to help her for a while until the "Most High" placed someone in her path. Christiana did not know what she was going to do. She also knew that she needed a job to be able to support her family.

Christiana would take her lunch each day and, at the same time, go and pick up her son, Cassius, from school. She would make her son snacks to keep him from being hungry until she was able to get back home to him. Christiana also explained to him not to answer the door if someone were to knock on it, and she would call him every five to ten minutes to talk to him, making sure that he was not scared.

Christiana made sure that movies were set up for her son in her room to watch on low volume, trying not to draw any kind of attention to her apartment. She made sure that all doors and windows were secure and made sure that he knew how to call her.

Each day, Christiana could not believe that her parents would not make any provisions to pick up their grandson, knowing that he was at home alone at such a young age. Christiana's friend, Kym, found out through a discussion about what was going on with Cassius being left alone for the last two weeks, learning that neither her parents nor Di'Ban cared to help with making sure he was safe.

Kym said, "Oh no, I will help you, Christiana, but it's a shame that your family is not trying to help and are okay with leaving their grandson alone, knowing that anything could happen. This is so selfish." Christiana was so appreciative of Kym's kindness and concern, which relieved a lot of stress off her. Christiana let Kym know how eternally grateful she was for the help.

Christiana's friend would drive from one side of town to the other to pick Cassius up. Kym would go and pick Niles up from the daycare, she would bring them back home, and she would not charge Christiana one red cent.

As soon as one hurdle was over, Christiana was met with another one when Niles had taken ill. One morning, Niles started vomiting shortly after she dropped him off at the daycare center. She was called to retrieve Niles. When his mother arrived, she could see that he was ill. Thinking it was probably nothing serious, Christiana took him to a sitter and then returned to work, thinking that it would pass soon.

Christiana kept calling the sitter to check on her son throughout the afternoon until her shift ended. Christiana called Kym, who was already looking after Cassius. She informed her that she would be home soon because she had to stop by a sitter to pick up Niles because he had taken ill at the daycare and that she didn't need to pick him up for her.

As soon as Christiana placed Niles in the car, he began crying and held his stomach. Christiana was trying to calm her son, letting Niles know that they were almost home and that she would take care of him.

Christiana became more concerned after she saw Niles lean over the armrest in the rear seat and immediately fell asleep. Christiana's intuition told her that something was seriously wrong. She rushed and pulled up to their home. She called Niles out, his eyes opened, but he then went right back to sleep.

Christiana picked Niles up. She carried him over her shoulder into the house, and when she laid him in his bed, she noticed that her son looked like a skeleton on his face. Christiana immediately called the children's hospital, where she took them for their health care.

The on-call doctor instantly began asking Christiana about the other children at the daycare center. She told him that she was not calling for anyone else's child but for her own child at that moment.

While Christiana was on the call, Niles woke from his sleep and then came to where his mother was. He was crying, and when Christian went to attend him, he laid over on her leg and went right back to sleep. Christiana was trying to wake Niles up, but it was a struggle to keep him awake. Meanwhile, Kym told her that she would keep Cassius. Christiana immediately picked her son up and rushed him to another hospital's emergency room.

Christiana called Di'Ban to let him know that she thought something was seriously wrong with Niles. Di'Ban told her that she was probably overreacting as usual. All she had left to say was, "I will call you and let you know what they find out."

The emergency room doctor stated to Christiana that they were having a hard time keeping Niles awake, and they also found blood in his stools. The doctor stated that they were immediately admitting him and would run more tests to figure out why Niles had internal bleeding.

Later in the evening, Christiana was informed that Niles had a tumor on his pancreas, which was causing the internal bleeding, and that he would be going into emergency surgery in a few hours.

Christiana called Di'Ban, informing him of what was going on with Niles. He then wanted to know when the surgery was scheduled. Christiana told him first thing early that morning, seeing that it was already

2 AM, and as soon the surgeon was to arrive, that was when it would happen. Christiana also told him that she would advise him to get there now because it could be sooner since it was considered an emergency surgery.

The pediatrician immediately made his way to the hospital to check on Niles before the surgery. He told Christiana that it was necessary. The pediatrician stated that he was also getting ready to contact the previous hospital regarding their conversation and to let them know the seriousness of Niles's condition. The pediatrician told the previous hospital heads that if his mother had not used her intuition a day or two later, her son most likely would not have survived.

Christiana stayed with Niles, never leaving his side, nervously waiting. However, Di'Ban didn't make his presence known until 7 AM, right before his son was being wheeled down for surgery. Di'Ban stayed until Niles was out of surgery but then left for the remainder of the day.

Christiana called to see if he was coming back to the hospital because she needed to go shower and change into fresh clothing. She also stated she needed to get clothes for the rest of Niles's hospital stay. Christiana had to have a family member go to her home to retrieve items that she needed because Di'Ban claimed he was too busy and that he could not be there to just sit at his son's bedside, stating that he had to work as if she did not have a job.

Chapter 38: The One Who Got Away

Christiana had gone to her parent's home one day for a visit right before she was about to move out of one apartment into another. While standing in the driveway, to Christiana's surprise, Kyrie was back in town. He pulled up on a motorcycle to visit her.

Christiana and Kyrie made their way inside the house to talk but mainly to be out of the range of those listening ears. Natalie was out front when he pulled up. She looked at him salty. One could see that she still held a grudge towards him. That was the first time Kyrie got the opportunity to see Cassius and Niles, but he then showed her a picture of his firstborn as well.

Kyrie asked Christiana what she was going to be doing later that evening because he wanted to spend some time with her. Christiana told him that she would be busy getting ready for her move the following day. He then wanted to know if he could come and see her after she was done moving. Christiana gave him her new address and then told him that he was welcome for a visit later the following day and that it would be nice to spend some time and catch up.

Kyrie showed up at her new place the next evening. They sat talking and reminiscing over nostalgic memories together. They shared fun laughs and decided to take a short car ride to get ice cream, then made their way back to the apartment right afterwards.

Apparently, those nostalgic memories must have been too vivid for Kyrie to handle because he suddenly needed to go. Kyrie got up from the floor, where they both were seated, and walked out. Christiana thought that he was just joking and probably was standing on the other side of the door, waiting for her to look for him.

Christiana remained seated on the floor for another couple of minutes and was expecting him to reenter, asking her something as simple as, "So you weren't really going to come and look for me?" Christiana finally got up and looked for him, but he had left. All she was left with was to say, "Well damn, I did not know that those memories possessed such power."

Chapter 39: Jealous Rage

Christiana, Di'Ban, and her childhood friend, Sicarri, and his girlfriend, Angie, had been introduced to each other. There were many times when they all hung out together. It was transparent that Christiana and Sicarri were seemingly more like a brother and sister until Di'Ban's delusional ego got in the way.

One evening, Christiana's friend, Sicarri, stopped by and brought dinner. Sicarri sat and ate dinner with her and the boys. A little later in the evening, after Sicarri left, Christiana and the boys were winding down for the night.

Di'Ban called, asking Christiana what she and the children were doing. Christiana said that she was cleaning up the kitchen from dinner, and the children were in the bathtub, getting ready for bed.

Christiana also mentioned that Sicarri stopped by and brought a movie that they all watched during dinner. Shortly after their conversation, Christiana heard a knock at her front door. It was Di'Ban. The kids were in bed by then, but Cassius was not asleep. Di'Ban didn't know that but he would find out later that his son heard him.

Di'Ban instantly started an argument with Christiana regarding Sicarri's visit. Di'Ban jealously and lividly said, "What the fuck was another mother fucker doing over here having dinner with my kids for?

Christiana said, "First off, this is not your place, and it's not like you don't know Sicarri." She further said, "Really, what you need to do is leave!" Christiana went and opened the door, asking him to leave her premises.

That was when he snatched the door from Christiana's hand, slamming it shut. He said, "I'll leave when I am good and got damn ready to leave!" The behavior that he displayed frightened Christiana because it caught her off guard, so she decided to remain quiet, hoping that he would just leave.

Di'Ban made up his mind that he was going to show Christiana that he somehow thought he owned her. Di'Ban saw that she was in a nightgown. That was when he pushed her over one of the dining room chairs, holding her down while he forcefully inserted his penis into her. However, Christiana fought back, getting him off her.

Di'Ban reached back and came forward with a closed fist, punching Christiana in her mouth. Di'Ban hit her as if he was hitting a man. The blow from that punch knocked her over a large chair onto the floor. Christiana felt her brain when it shook from the force of the blow that he had inflicted upon her.

Christiana's mouth instantly felt full of fluid, and blood flowed down the front of her gown once she reached to touch her mouth. It took Christiana a few minutes to gather herself, but once she got up from the floor, she noticed that Di'Ban had already cut on the dining room light, and that was when he saw the damage that he had caused to Christiana's mouth.

Cassius opened the bedroom door and called for his mother as he heard the struggle. She told her son to go to bed. He told his mother that he had heard the scuffle that had caused him to come to his bedroom door. Christiana told Di'Ban that if he didn't want his children to see him getting arrested, she would advise him to get the hell out right then.

Di'Ban, after seeing what he had done, tried playing an empathic role by saying, "I'm sorry. I don't know what I was thinking. That is not me! Christiana, please forgive me. Do you want me to take you to the hospital?" Christiana said, "You are sick. First, you kick my ass, then you want to take me to the hospital? Just get the fuck out!"

Chapter 40: Aftermath

After that abusive episode with Di'Ban, Christiana called her close neighbor Naomi to help her get herself together. Christiana was in a state of shock from what she had just endured at Di'Ban's hands.

Naomi arrived and was shocked at how busted Christiana's mouth was. Naomi had to take scissors to cut the excess skin that was hanging from Christiana's mouth. Naomi then helped clean up the blood and got Christiana into some clean clothing. She then asked, "What the hell was Di'Ban doing and thinking? I know that he's met Sicarri before, and as a matter of fact, you all have been to many outings together. Damn, is he that fucking insecure with his bitch ass!" Christiana had fallen to her lowest state of mind after what had happened to her. She did not even want her family to know about this episode.

Many employees asked Christiana about what happened to her mouth. Christiana lied and stated that she got hit in the mouth by one of the kids at her son's pee wee league games with a bat.

Christiana still carried on as if nothing had happened, or at least on the outside, it seemed, but that inside was a raging inferno waiting to explode. Christiana still dealt with his sorry ass and with much more of his bullshit for a while. However, it was going to get worse before it was going to get better!

Christiana had a friend, MyKale, who had known Christiana for a long time since they were kids and who did care for her. MyKale often asked her about her relationship with Di'Ban. One day, as they were discussing Di'Ban, he told her that he did not trust him, and he meant that in the worst way.

MyKale decided to give Christiana something to protect herself against Di'Ban after she told him about what he had done to her. He gave Christiana a military-grade Kaybar combat knife and then taught her how to use it. One evening, MyKale stopped by to check up on Christiana. While they were standing out conversating, Di'Ban pulled up and jumped out of his car. He ran up to where MyKale and Christiana were standing, looking foolish. He told MyKale, "Let me tell you about this bitch!"

Christiana looked at him with nothing but disgust, but what he didn't know was that Christiana had recently discovered that he had been sleeping with his brother's girlfriend. Christiana looked at Di'Ban and then said, "You can stand your ass out here, all you please, to run me down, but just know that I don't give a damn, just don't bring the shit to my door!"

Christiana told MyKale, "I'm going inside because I need to feed my kids, and I need to get ready for work. Now you can stand out here to listen to whatever nonsense his ass wants to talk about, but I have other things to attend to."

MyKale called Christiana afterwards, laughing, saying he kept him out there sitting on his car, snotting and lying for an hour and a half! MyKale also said, "Listen, all he done was bitched and moaned and, oh, yeah, while lying about how much money he claimed that he had spent on you!"

MyKale said, "The only reason that I stayed and listened to his sorry ass was that he's older and out of respect for my elders," while laughing. MyKale said, "I wanted to tell his ass if that was all that you've spent on her, then you are one cheap bastard because, after all, your sad ass has kids with her!"

A few months later, Christiana went for her yearly checkup. A day later, the doctor called her to inform her that her yearly gynecologic test results were fine but that she was also pregnant. Christiana could not believe that she was pregnant again. She asked the doctor how it was that she could be pregnant because she had been using her diaphragm. The doctor told her that it was sized to fit but that he was uncertain of what had happened and that she should examine it. Christiana went home to examine her diaphragm, and to her surprise, some holes had been punctured, which she would not have seen without closely examining it.

Christiana told Di'Ban that she was pregnant, and as usual, he caused chaos, upsetting her. She told him that she was not feeling well at the end of his rant and that she was going to leave the boys with him and was going home after that,

Christiana called to talk to MyKale regarding Di'Ban's nasty disposition when she told him about the pregnancy. While they were on the phone, Christiana asked MyKale to hold on while she ran to the restroom; that was when she saw bright red blood, and she panicked.

Christiana went back to her phone call with MyKale. She told him that she was bleeding, and that was when he asked her if it was bright red and if it was that she was most likely having a miscarriage. MyKale was a medic in the military, and he stated he had witnessed miscarriages before.

Christiana went to the hospital the following day and was told that she was going to miscarry because her beta count was not where it needed to be at that time. Christiana was then scheduled to go back the following morning to have a D and C performed for her miscarriage.

Di'Ban had stolen the code on her phone when he was sneaking and entering the code to listen to her conversations. He recorded them to be able to use later in life to convince his children that the conversations were something other than what it was. He was a person with a hardened, ugly heart, and he could not handle loss even if it was by his own hand, and he needed to taint his sons' minds about their mother for some get back.

Twelve years later, that stale, dysfunctional relationship continued, and again, Christiana got pregnant. Meanwhile, Di'Ban continuously played the role of the weekend father at Christiana's" home. This pregnancy brought a challenge for the baby and for Christiana that caused her to be placed on bed rest until the birth of the baby.

Christiana told her mother, Natalie, over a phone conversation that she did not know what Di'Ban was waiting on to get the baby a crib. That was when Natalie said, "Well he told me that he doesn't even think that the baby was his!"

Christiana asked her mother, "So when were you going to tell me what he said? I mean, since you both are so comfortable talking about me together?" Christiana said, "No, wait, Mother, please tell me, how long have you known about this?" Natalie said, "He told me several months ago when he stopped by one day." Christiana already understood that they were both two of a kind and hearing how she was fending for him.

Christiana confronted Di'Ban when he showed up later that evening at her home. Once he walked in, feeling all confident, that was when Christiana said, "Now you know what's so amazing?" Di'Ban looked clueless. Christiana said, "It's how you keep claiming kids aren't yours, that you keep managing to pop out of your dusty ass nut sacks!" Christiana continued and said, "I tell you what! I just found out today you've denied this baby. Please take your ass with your dusty nuts and get out of my house!"

Two months later, Christiana went into labor, and she delivered another son whom she named Ishmael. Christiana did not even bother calling Di'Ban to let him know that he had another son. Di'Ban was allowed to pick up his other two sons without coming into her home. However, she told the kids not to mention that she had the baby. However, they could only tell if their father asked if she had the baby then they could tell him. It was a month after the birth of Ishmael that Di'Ban finally asked Cassius and Niles if their mother had the baby, which they informed him that she had.

Christiana received a phone call from Di'Ban one evening. He said, "I heard you had the baby!" Christiana stated, "I had my son." He then asked, "How come you didn't call me?" Christiana said, "Call you for what? Remember you said he wasn't yours."

Di'Ban ignored her question. He then asked how the baby was doing and what she named him. Christiana said, "I named him Ishmael, which I took from the Bible, where Hagar and Ishmael were set aside in the desert by his father." Christiana had to ask if he was still there because she knew he had no clue as to what she was even talking about.

Di'Ban asked if he could come to see Ishmael. Christiana asked him, "Are you admitting that he is your son? Otherwise, there's no need for you to come to see him." Di'Ban said, "I know that he is my son." He then showed up shortly afterwards to meet his son, but truthfully, he only wanted to see if Ishmael looked like him, and to his surprise, Ishmael looked exactly like him. Christiana was ignorant to the fact that Di'Ban had already told everyone that Ishmael was a fireman's baby, but it wouldn't be until later that Ishmael would also learn from his brothers what his father said about him.

Christiana's survival techniques kicked in for taking care of her children's needs. She returned to work when Ishmael was only two weeks old. Christiana was fortunate to be working for a group of men who were business partners that consisted of Doctors and Lawyers.

Christiana kept money because the men whom she worked for, paid her well. They made sure that she and her kids needed nothing and saw that she was a hardworking mother. Christiana was introduced to an elite crowd of people while working for them.

Di'Ban had become envious of how well Christiana was maintaining, and she was not looking back. Christiana was at work one evening at one of the lawyers' offices when she had to take the kids with her. It made it difficult for her to get any work done with the children. She called Di'Ban and asked if he could come to get the children so that she could continue to work.

Di'Ban came to pick the children up. He then took the kids to the car but returned knocking on the office door to be nosey. Christiana answered the door to see what he wanted. All he wanted was to question her about who was there with her.

Christiana told him that she did not have time for his nonsense and then asked him to leave. As soon as he left, the lawyer came out of his office to check on Christiana. He then asked her if he needed to speak with Di'Ban about harassing her. Christiana told the lawyer that she would handle it and it would not happen again. She made sure to secure a sitter the following day because she saw that he was attempting to mess up what she had going.

A few months later, an event took place, which Christiana was also a part of since she was employed by this group of businessmen. The event consisted of some of the top lawyers and doctors in Chicago. One of the men that Christiana worked for told her that there was an influential lawyer who had been inquiring about her. This partner also mentioned that this certain lawyer would probably approach her at the upcoming event so that she would not be caught off guard.

Once Christiana arrived at the event, she took a seat with those partners and their wives. Shortly after the event started, one of the partners looked up and saw Di'Ban standing in the entryway of the ballroom.

Christiana saw that he was not alone but was with his buddy Diego, the same guy who attempted to hit on her.

That same partner, who had already seen Di'Ban before, asked Christiana if she recognized who that was at the entryway, she responded by saying, "Yeah, that's his sorry ass!" Di'Ban and Diego made their way into the event and then came up to the table where Christiana was sitting. Di'Ban asked Christiana if she was having a good time. Christiana responded that she was having a great time. Without her even inquiring about what he thought of the event, he stated that it was "nice and all," but it was not his crowd-type. Immediately following his first question, he then asked Christiana about where the children were.

Di'Ban was trying to let everyone at the table know how he was tied to her. That was when that influential lawyer that Christiana was told about walked up with what seemed like perfect timing. This lawyer grabbed a seat and pulled it up right next to Christiana. The lawyer showed Di'Ban that he did not care about what or why he was standing there.

Diego decided that he would puff up both their egos for the men at the table. He said, "I told Di'Ban that you are the most beautiful woman here in this room!" Everyone at the table was looking for Christiana's response like it was a tennis match. Christiana looked at them both, and without missing a beat, she said, "The table with the food is behind me. Please enjoy yourselves!"

All the partners at the table looked at each other and chuckled. The influential lawyer immediately sat back in his seat. He looked at Christiana with wide eyes and smiled as if to say, 'Ma'am, you handled that with class.'

The influential lawyer asked one of the partners while Di'Ban and Diego had not yet walked away, "Is anyone going to introduce me to this beautiful woman?" The men damn near lost it, telling him, "Man, now you know damn well that was uncalled for!" The influential lawyer responded by briskly stating, "What? They better look me up."

Di'Ban and Diego had no response or recourse for anyone at the table except to come and walk back through with women who looked like they were picked up off some solicited street corner. Di'Ban was attempting to get a reaction out of Christiana, but right at that moment, one of those lawyers who saw these women leaned over to Christiana and said, "All I can say is, they must have skills because damn!"

Christiana was tickled by his comment and said, "I never thought I would really get a chance to actually view some alley cat up that close!" Christiana went over into the room where the DJ was playing and then hit the dance floor with that influential lawyer. Then came Di'Ban with one of the women who was holding onto his arm like he was the man of the hour.

Di'Ban and his crew took a seat near the dance floor where they could watch Christiana. Once she noticed them watching her, she decided to leave because the sight of Di'Ban literally sickened her.

Christiana eventually left the event. She went to pick her children up. As soon as she headed home, she saw Di'Ban stalking behind and passing her. Di'Ban also saw Christiana's car, so he went right back, hoping to have a conversation with her, but Christiana was not having it, and sent him on his way.

Chapter 41: The Plot Thickens

Di'Ban could tell that Christiana was changing into a new woman. It seemed to entice his curiosity. He understood that he had to come up with some new strategies to gain her back.

One day, Di'Ban called and asked Christiana if he could come to talk with her. She was skeptical of what he could possibly want to talk to her about. They met up at a coffee shop, and once seated, Di'Ban told Christiana that the Most High came to him in a dream and showed him his ways.

Di'Ban went on to say that he was also sorry, for he knew how he had treated her. He asked if she would please give him a chance to show her how much he loved her. Di'Ban said that he needed her because he knew she was a spiritual woman and that he was ready to settle down. He then asked Christiana if she would marry him.

Di'Ban stooped as low as the pit of hell by using the Most High's name for his scheme to soften her heart because he knew that she was spiritual. Once more, Christiana fell for the lies, believing that he was a changed man because it was one scheme that he had never used before.

Christiana went to inform her parents that she and Di'Ban were going to get married. They did not show any interest in helping her. Christiana found a beautiful wedding dress, and she then asked Nadia to be her matron of honor. Christiana invited all her girls to be part of the ceremony to come and try on dresses she chose for them.

Christiana asked the saleswoman who was attending to her to put her dress away because she did not want anyone to see it. While she was working with her bridesmaids, Nadia saw the room that her sister came out of when she first arrived. She figured that her dress must be in the dressing room because she had never seen the saleswoman come out with anything.

Nadia then snuck into that room to see Christiana's wedding dress. Once she saw how beautiful Christiana's dress was, she became jealous. Nadia knew her sister had exquisite taste, but she couldn't believe that her sister would spend that type of money. When she read the price tag, she realized that she had only spent fifty dollars on the dress for her and Damon's ceremony.

Christiana was so busy attending to her bridesmaids that she didn't notice that Nadia had slipped away or had seen her gown. Christiana went to talk to her mother to see when they could get together and look for her dress for the wedding. That was when Natalie informed her daughter that Nadia and she were going to go together to look for her mother's dress. Christiana could not understand how her mother could be so inconsiderate even when it was supposed to be a shared moment.

Natalie then offered her opinion about Christiana's dress. She said, "I can't believe that you are going to spend that amount of money on a wedding dress." Christiana asked, "How did you know how much my dress cost?" Natalie did not hesitate and stated that Nadia had told her. Natalie tried to make her feel as if she was being outrageous about spending that kind of money on her dress!

Natalie said, "Well, Nadia and I only spent fifty dollars on our dresses!"

Christiana said that was good, and she wished she could have found a dress she liked and it only cost fifty dollars. But she had not asked anyone to help her pay for her dress, so it shouldn't matter how much she was spending on it.

A few days later, Bash called Christiana and said, "You are going to have to change your wedding date." Christiana asked her father, for what reason do I need to change the date because we are only a couple of weeks away?

Bash said, "I have an event that I want to attend." Christiana said, "OK, but I'm not changing my wedding date. However, you can go to your event, and I'll find someone else to give me away!" Christiana said, "Actually, I'll have one of the guys who I work for give me away because they would be honored to do it for me!"

Christiana was putting a beautiful wedding together with no help from her mother or her sister. She was very capable of pulling things together in such a lavish way, which was something that Natalie and Nadia both disliked about Christiana. Whatever she puts her hands on comes out breathtaking.

Christiana was only supported by her friends, but without any help from her family, she wanted to show her appreciation to those people who helped her. Christiana prepared a lavish hostess brunch for them, with gift bags for each one of them. They were very appreciative of her gesture towards them.

The wedding day was soon approaching when Di'Ban decided to blindside Christiana with a prenup that he had drawn up and presented while they were on their way to pick up some last-minute items for the wedding.

Di'Ban pulled the prenup out and handed it to Christiana. He said, "I need you to sign this."

Christiana asked, "You need me to sign what?" Once she looked and saw it was a prenup, Christiana chuckled, but her soul laughed. She eventually said, "You got to be out of your damn mind!"

The fact that Christiana chuckled left him speechless. Christiana said, "I'm not signing anything, and besides that, you don't have anything that I can't achieve on my own! Now, here's what you can do with that prenup. Please stick it up your ass. Oh, but trust me, we got time to call this wedding off! I am not above sticking a sign on the front door of that church letting everyone know that this wedding has been canceled, so what do you want to do?"

The three nights before the wedding, all the wedding parties were given a specific time to be at the church for rehearsal. The only people who showed up late were Bash, Natalie, and Nadia, and Andre did not even bother to come. The family walked in one hour and a half past the scheduled rehearsal time with shit-eating grins on their faces.

They were hoping to see if they had upset Christiana, but they immediately claimed that she had given them a different time. Christiana said that couldn't be the case because everyone else was on time. She decided she was not going to argue about it.

Christiana's family saw how beautiful everything was for the wedding and looked with disbelief at what she was able to accomplish without them. What was even more disheartening was that her parents did not attempt nor offer a dinner for the wedding party or anything else.

Nadia and Damon were very rude and disrespectful as they asked Christiana if she would have cable set up at the reception for a fight party. Christiana told them that if they wanted to watch a fight, they were both able to leave the reception and take it to their home for a fight party. Christiana made sure to let them know that it was not about what they all wanted and not to disrespect the reception.

Christiana called Di'Ban early in the morning before they headed to the church. She asked him if they still wanted to go through with the ceremony. Christiana could not get over how he blindsided her the day before, but she reminded him that they could call it off. Di'Ban just stated, "I will see you at the altar," only because he still had a financial plan.

Once Christiana arrived at the church to start getting dressed for the ceremony, neither Nadia, her matron of honor, nor her mother came to assist her with getting dressed. Christiana was in a room all by herself, attempting to get dressed until one of her hostesses took notice of them not helping her. She then came to assist Christiana to help her get dressed. This hostess was in total unrest with Nadia's behavior. She stated, "Your sister is not shit for not helping you, but I've got you, and do not let them upset you today because I can see that they are on some foolishness."

Once everyone got into place, the wedding coordinator came to Christiana and told her how beautiful she looked, and they were ready to start the wedding. Christiana came out of the dressing room. There stood her father, Bash. He sarcastically looked at Christiana and said, "I might say I don't."

That is when the coordinator said, "No, you are not. This is her day!" Apparently, what the coordinator said to Christiana's father made him feel shameful because he walked his daughter down the aisle without mishap.

When the church doors finally opened to walk her in, she saw Andre up in the pulpit, walking around as the ceremony had already begun before he took a seat. Once Bash gave his daughter away and the preacher began speaking, Nadia kept turning around to talk to Natalie as if the ceremony was not even happening. Christiana also saw that Di'Ban did not have any guests on his side of the ceremony. That proved he had not sent out any of his invitations. Shortly into the ceremony, Christiana sang at the wedding, which was something most of her guests did not realize she could do. That was when Andre leaned between the aisle, having a full conversation with Bash.

After the wedding ceremony was over and everyone made their way to the wedding reception, their guests were ultimately blown away at how over the top and beautiful the reception hall was. Everyone at the reception wanted to know who decorated the reception, and all they heard from the bridal party was, "By no other than Christiana!" Christiana's guests approached her before they left to offer an apology for the way her family had acted, stating that they could not believe their behavior. The DJ asked Di'Ban, "Where are the single ladies at?" doing their first dance. Christiana heard when Di'Ban said to the DJ that they were taking applications. She found that comment to be disrespectfully trifling.

Unfortunately for Christiana, her husband had not made any attempt for a honeymoon. What they had the following morning for a honeymoon was nothing out of the ordinary of a pancake breakfast.

Chapter 42: Money Was the Root

While institutionalized, Bash's sister Angela ended up getting pregnant and having two children, Cynthia and Shelia, by either patients or staff members who took advantage of her. After 25 years of being institutionalized, Angela was able to leave the mental hospital and live on her own with her daughters, which was a struggle. However, she did her best to live a somewhat normal life with the skills she was equipped with.

The children also were born with mental issues. Bash had received a call from some of his sister's church members one day, letting him know that his sister Angela had suddenly died. The church members took care of Angela's daughters until the family came to their rescue and made provisions for their lives.

Unfortunately, Gia had grown older and was unable to tend to the needs of Angela's daughters. So, Bash and Natalie decided that they would bring them to their home. However, it was only for the monthly disability checks that the girls would bring with them.

One of the girls made the decision that she wanted to stay with her church members, while the other girl was not able to lend her opinion as to where she wanted to stay. So, she was brought to Bash and Natale. Sheila was in for one hell of an ordeal, living with Bash and Natalie. She was treated like an unwanted guest in their home. Sheila could not go to the restroom to shower in peace without Natalie terrorizing her by banging on the door, rushing her out, and then suggesting that she had hygiene issues. Natalie told Sheila that she needed some bleach in her bathwater because she smelled.

Shelia was in the restroom, bathing one afternoon, when Bash started banging on the bathroom door, telling her he needed to use the restroom. When Shelia didn't come out fast enough, Natalie screamed for her to come out right away.

Sheila was nervous from all the yelling. She came out of the bathroom butt naked with soap still on her body. She went into another room and sat down on a chair, waiting to return to her bath. Natalie came into that room yelling at her again because she was sitting on a chair wet.

Natalie constantly would get on the phone, calling different family members to talk about Sheila about anything to anyone who would listen. Whenever she would call and talk to Christiana complaining about Sheila, Christiana often told her mother that they needed to stop yelling at her because she was adjusting to the loss of her mother, and it was doing nothing but making her nervous. Natalie would respond to the comment by saying, "You don't know the shit she does with her funky ass," attempting to convince Christiana.

Christiana started to take Sheila on the weekends when she could, pretending to give her parents a break, but in reality, it was to provide Sheila a break from them.

Chapter 43: Apples Don't Fall

Andre invited the family over for a BBQ at his home. Christiana went, and she brought Sheila with her. They were eating and enjoying some music with Andre's girlfriend, Monica. Andre showed up late for his own BBQ. Andre was intoxicated as usual. He began antagonizing Shelia for his own enjoyment and entertainment purposes. He started saying things that he should not have been saying to her, asking her un-appropriate questions like, "And what's up with your period, damn?"

Andre had been listening to his mother as she tore down Sheila daily, which made her life a living hell. Natalie treated her like she was a dog, and so did the other family members. Andre would not let up on Sheila, but that was when Christiana stood up for her. She said, "You are going to leave her alone because Shelia is sitting here minding her business, not messing with anyone. Now, here you come trying to make her feel less than, but it really says a lot about you!"

That was when Andre pulled a gun out of his pocket to show Christiana or strike fear in her. However, she was not afraid, nor did she back down from him. What she told him was he was a small man for attacking Sheila and picking on someone who couldn't defend themselves, and if he was going to pull a gun, he should be prepared to use it.

Christiana had gone over to her parents the following week after the BBQ. The conversation came up about how Andre had treated Sheila. Christiana told Bash how Andre acted towards her and the things he said to Sheila. Christiana had also told her father that Andre pulled a gun on her for defending Sheila.

Bash began making excuses in the defense of Andre, stating that he did not mean it. Christiana said he knew exactly what he was doing, and that watered-down excuse they all were making, they could miss her with it, and there was no way she was just supposed to accept it like the kiss of Judas.

Chapter 44: The Beginning of The End

Christiana had already given a move-out notice to the place where she was residing with the children right after the wedding. She asked Di'Ban to take the time to make room for her and the children so that they could begin moving and living as husband and wife should.

Six weeks later, Di'Ban had not moved even one box out of the way to make room for his family. However, he saw that Christiana was already packing up for the move. Di'Ban told Christiana that he needed more time because his schedule was tight.

Christiana told him that they did not have more time because she had scheduled the move-out date. Christiana had gone over to the house one day, and what Di'Ban did not know was that Christiana had seen him when they walked in, where he kicked his caller ID box underneath the couch.

Christiana told Di'Ban that she was taking Cassius on a scheduled trip to DC the following week, and when she would return, he needed to be ready for them to move in. Christiana did not forget what she had seen with the caller ID box. She was going to ask about it as soon as she arrived back in the city.

Christiana got back from the trip, and when she went to his house, she went through the caller ID. What she found on that caller ID box was surprising because it was the woman's name, Alley, whom she had already heard about. Alley, who happened to be James's, his brother's, girlfriend.

Christiana jotted down Alley's full name and her phone number before Di'Ban arrived back into the house from taking out the trash. Christiana went to work the following day. She called one of her friends who had the capability of finding out where anyone lived.

Christiana also talked with one of the businessmen, who was a lawyer, and told him that she believed that Di'Ban was having an affair. The lawyer said, "No, you all just got married." He asked Christiana what she wanted to do once she found out if it was true.

Christiana said, "I need to see this for myself because he will not be able to use any of the excuses." She said, "Then I am going to file for a divorce because I have wasted too much youthful time of my life with him as it is!"

Christiana received Alley's address right away from her friend, and she and the lawyer took a walk to where you can locate anyone's house on a map. The lawyer showed her exactly where the house was. He then told Christiana not to go there at night because it was a strange neighborhood but to go during the day to locate the house while everyone was at work.

Christian followed the lawyer's advice and then drove to the location of Alley's house which happened to look more like a shack to her. There were some guys standing out front across the street from Alley's house.

Christiana talked with the guys standing out across the street for a few minutes. She stated to them that she had not been over there in a long time and that she was looking for Alley, stating she believed Alley lived there and that it had just been a while since she had been to her house.

The guys told Christiana that she did live there. Christiana thanked them. She very politely said, “I’ll just come back later once I think she’s home.” Christiana asked the guys, “What time does Alley usually get home, and what kind of car did she drive?” To her surprise, they provided all the information to her questions.

Christiana did go back later that same week. She then found a great stake-out spot and watched Alley’s house without ever being noticed. Christiana parked way back up in the dark. Her windows were tinted so no one could see that she was in the car.

Christiana studied her moves. She could even tell what room Alley slept in just from watching. She knew that only a certain bedroom light would often go on and off. However, she hadn’t figured out if Di’Ban was still seeing her. After the second week, she thought that she was wasting her time with the surveillance, but Christiana would soon come to learn that all she needed to do was to listen.

Christiana was having dinner with Di’Ban one evening while they were watching a movie. It was a movie that exposed where the husband was cheating on his wife. Christiana smoothly turned and looked directly at Di’Ban, then asked him in a hypothetical tone, “Why do people get married just to go and cheat?”

Di’Ban evidently felt disoriented after the question that she asked and did not realize that his abrasive response was seen as a guilt-like answer. He said, "How the fuck would I know!" The response was all Christiana needed to hear, as it caused her to keep following through with her surveillance of him and Alley.

Chapter 45: Ready or Not

Christiana went and spoke with the leasing manager of the apartments where she lived to find out if she could remain in her place. Unfortunately, she was informed that they had the apartment already scheduled for another tenant to move in.

Christiana told Di'Ban that he needed to help her secure another means of housing for her and the kids, or he was going to open his door to them. She said either it was going to be willfully or unwillingly, but if he chose the latter, then she would gain entry by any means necessary.

Meanwhile, one evening, Di'Ban had stopped by Christiana's apartment when the older boys asked if they could go with their father to spend the night. Di'Ban instantly made excuses to the boys, stating that he had an early morning appointment. He told the boys that he was on his way home to go to bed and that they could not spend the night because he had to rise early the following morning.

Christiana sat on the couch as if she was engrossed in what she was watching on the television while putting Ishmael to sleep. She made sure that he would not see that she was paying attention to what he was saying to the other children. Di'Ban believed that he knew Christiana like clockwork because she lived a structured life. He also knew that she would be putting everyone down for the night, including herself, shortly after he left.

Christiana waited until it was well past her bedtime, giving Di'Ban time to call to see if she was asleep and would answer the phone, which was something that he usually did. Christiana realized what he wanted to make sure of was if the coast was clear enough for him to comfortably creep out. It was just as Christiana had predicted, and once he called Christiana, she played her role well, sounding as if she was asleep. Di'Ban said he didn't want anything and that he was just checking to see if the kids were upset with him, but once she told him that they were alright, Di'Ban immediately let her go. But to his surprise, Christiana had already called her a sitter, and she was going to leave to check out his story.

Christiana dissected his routine after he told the kids no; then she thought of when he came and stayed with her and the kids, and she realized that he was cheating through certain times of the work week all because he would not let the boys spend the night with him. That was why she chose to get a babysitter that evening.

Christiana first drove past Di'Ban's house. That was when she saw that he was not home and that he had not even gone home after he left her home. She knew that he had not gone home because the porch light wasn't even on. Christiana sat there for a minute contemplating, making the drive to Alley's house. That was when the right side of her brain told her to go, but the left side said, 'You have been over there for the past two weeks, and you have not seen anything!'

Christiana listened to the right side of her brain. She popped in her favorite driving music CD and cruised on her way to Alley's house. Once Christiana turned onto Alley's street, there sat Di'Ban's vehicle.

Christiana felt her adrenaline pumping, and she was spazzing out. She realized how much she was spazzing because when she looked in the rearview mirror, she saw that she was talking to herself, saying all sorts of unbelievable comments.

Christiana saw that both cars were there. That was when she cut her lights out and crept up behind Di'Ban's vehicle. She got out quietly, then walked up to the house and politely knocked on the door. Christiana saw there was only a flicker of a flame that was burning. She thought, Oh, they up in here fucking. There was a barking dog that had come to the front door, and Christiana thought, I guess he's use to this motherfucker as well, while softly chuckling.

Christiana did not want to upset their night, but Di'Ban was going to face her one way or another. She gave another polite knock. The next thing that Christiana heard from the window, which she had identified as Alley's bedroom, was Di'Ban's voice.

Di'Ban was shocked when he saw that it was Christiana. She heard him when he said, "Oh shit." However, still, no one came to the door. That was when Christiana said, "You might as well bring your ass to this door because you were a man when you brought your ass over here, so don't act like no pussy now because I'm not going anywhere!"

When she did not get a response, Christiana said, "OK, I see this is how you want to play it!" Christiana went to her car and got her Kay-Bar. She then put both their vehicles in an incapacitated state. Christiana figured that they could try to run if they wanted, but these bitches weren't going to get very far.

The next move was when she went back to her car and called Alley's house. Apparently, they were too scared to even answer the phone, and once it went to voicemail, Christiana sarcastically said, "I know you two motherfuckers are scared, I know!" Christiana went on to say, "Listen, y'all might want to call the police because you all are going to need them!"

Christiana got back out of the car, then she stayed on the sidewalk near the cars, calling for Di'Ban, asking him, "Are you coming out yet?" When there was no response to her question, she said, "OK, we are getting ready to see about that now. Bring your ass on out!"

Christiana started by knocking out their headlights before she asked again if he was coming out. When he didn't answer or come out, she kept knocking out different parts of both their vehicles, such as tires, windows, taillights, and side mirrors, and fucking up their paint jobs each time she asked if he was coming out.

Alley's neighbors were enjoying the entertainment. The porch lights were coming on left and right, which did not bring down Christiana's rage because of the level of disrespect for each of them. There just happened to be a policeman in the area. He showed up at the scene, and once he walked up to Christiana, he requested her to drop her weapon. The police asked her what was going on. She explained that her husband was in the house with another woman, and all she wanted to do was to talk with him. He then asked which house it was. Christiana was calm and cool while talking with one of Chicago's finest. He asked again where he was, and she directed him to the house.

Apparently, the police officer noticed how attractive Christiana was because he was looking at her deeply. Then he immediately went up to the house and knocked on the front door, stating that it was the police. It was visible that they were too scared to open the door. The policeman then went to the backdoor, where Di'Ban allowed him entry. However, the policeman brought him out of the front door while Di'Ban looked scared, shocked, and shaken.

The policeman must have seen Alley because when they arrived back to Christiana, the policeman asked Di'Ban in a disbelieving tone. He questioned him, "This is your wife?" The policeman looked at Di'Ban, then said, "Man, you need to get your shit together," as he shook his head.

Di'Ban instantly started to try and lie his way out of being caught, but that was when the policeman shut him down by saying, "Man, look, you got some serious damages out here!" The police went on to say, "It is my best advice that you need to call your insurance companies and tell them that you all don't know what happened."

The police officer said, "Now these damages need to be cleaned up," before he looked at Christiana and stated that he was getting ready to leave. The officer then looked at Di'Ban and said, "I'm returning in 30 minutes. If not resolved, then everyone is going to jail!"

Di'Ban foolishly looked at Christiana and asked, "How am I supposed to pay for this?"

Christiana said, "I don't know. Why are you asking me? But here's an idea. You better get your girl to pay for it!" Christiana got back into her car and told the police officer, "Thank you." She saw and heard that he did not want to arrest her. Right before Christiana left, she said, "Your divorce papers are coming soon."

Christiana and the children needed to be out of their place the following week, and Di'Ban hadn't made any provisions to help secure a place for his family to live. Christiana had also begun to sell many of her home furnishings because she wasn't aware of where they were going to be living.

A week later, Christiana called Di'Ban and told him that the next day, they would be moving in because they still had not secured a place for her and the kids. However, he tried to give her some push back, and she again said, "Like I said, it can be willfully or unwillingly, but we will be coming tomorrow!" Christiana also did as she had promised him the night when he was caught red-handed, she went and filed for a divorce in the same week after catching him cheating.

The following day, after telling Di'Ban that she and the kids were moving in, Christiana called her nephews, Ma'nias and Ya'fam, to help her with loading their belongings to move into Di'Ban's house. Once they arrived at Di'Ban's house, he saw he had no choice but to let his family in; however, he was very shitty, acting and throwing their things around in the house. Di'Ban was making sure to show his family that he did not want them there. Christiana's nephews and the two oldest boys had finally finished unloading the truck, and that was when Di'Ban" could no longer hold his emotions together. He loudly said, "Hell naw, I don't want yawl here!"

Christiana ignored his comment, knowing he did not care where they went or if they even had a place to live. She then began setting up each bedroom. According to the sleeping arrangements she made, Cassius got a room to himself. She then set the room up for Niles and their father to share, and then, finally, a room for Ishmael and herself to share.

Di'Ban set and calculated other plans that he was going to inflict to force Christiana and his children out of his house. He was consumed with himself, not caring if his own seed would be affected by his actions.

The first thing that Di'Ban chose not to dispose of was any of his luxuries. He went as far as setting an entire living room in the center of the kitchen floor to make them uncomfortable. However, Christiana and the kids just walked around it as if it was not even sitting there. When he saw that he was not getting the response he wanted, he conjured up other plans, attempting to make things worse.

One morning, right after Christiana had gotten Cassius and Niles off to school, she made sure she took care of Ishmael by getting him some breakfast and then setting him up for his cartoons to finish getting their belongings in place. Christiana started unpacking the children's clothing and putting them away. However, this was the morning that Di'Ban would attempt to reign his terror. Once he heard Christiana in the living room, ironing and trying to get things done, he decided to cock a shotgun from the other room for no apparent reason hoping to scare her. One thing that he did not understand was that Christiana had been raised around guns because her father was a man of many hobbies and one of those hobbies involved gun collecting; therefore, she was knowledgeable of the sounds of different guns. Her father taught his children how to clean guns, which was how she learned the different sounds of guns being cocked.

Her father showed his daughter how it took damn near a sniper's skilled ability to be able to hit moving targets; therefore, she did not fear guns because she understood the skill that came with using them.

Christiana was ironing, but once she heard the gun cock, she put the iron on high. She then immediately snatched that iron cord out of the wall. When she heard him cock the gun for a second time while listening for him to take any footsteps, Christiana said without hesitation, "You can cock all the shotguns you want, buddy, but you better just hope that you can get to me before I get to you!" Di'Ban heard her when she dialed that iron up and before she snatched that cord out of the wall. He was wondering if she was waiting for him to make a move. Christiana stood in defense mode, ready with that hot iron in hand, holding tightly on the handle while waiting to hear him take one step towards the door. She also understood that he was listening for a frightened response from her. That was when she said, "I hope you're ready to get your ass beaten with this hot iron if you come for me!"

Di'Ban heard no fear coming from Christiana. He immediately understood that he should put his gun away while hoping that she didn't come at him. Di'Ban also understood that she was not playing with what she had said to him and also knew his cowardly behind wasn't ready to have his skin taken off by a scorching hot iron.

Di'Ban didn't even attempt to come out of the bedroom for a while after what she had said to him because he remembered how surgical she was when she dismantled his and Alley's vehicles without even getting cut. However, he finally mustered up some nerve and then said, "You are crazy. Who thinks to do some shit like that!" Christiana said, "Say what you want, but I bet you are going to keep your ass in that bedroom!"

The next wicked demise he decided upon was to unhook the gas stove so that Christiana could not prepare food to even feed the children, which meant she was forced to feed the kids fast food for breakfast and for dinner most days. Christiana, at times, would buy food and cook it at someone else's house just to feed her kids.

The house windows were sealed down, but it did have central air, and it was in the midst of summer when the temperatures were in the high nineties. Di'Ban did not care about what he was going to do next, so he went and unhooked the central air passage. That coward then left his family to fend for themselves, but again, the concerns of comfort were only for himself. Meanwhile, Di'Ban took and stayed at Alley's every night to enjoy the comforts of cool air conditioning, while Christiana and the children were left to sleep with the front and the back doors wide open to breathe while using fans to circulate the air.

One of the neighbor's young sons, Victor, saw how Di'Ban was maliciously acting towards his family while leaving them in a dangerous situation. Victor made it a point to bring his dog, which was a two-year-old German Shepherd named Maximillia, to Christiana. He said, "I need you to keep this dog in the house for the safety of you and the boys."

Victor also stated that the son of a bitch is going to bust hell wide open because he did not even care about what could happen to his family, but she will protect you and the kids from someone coming in on you all through the night. Di'Ban saw that she had a dog for their protection and complained about the dog being in his house, but she told him that she was not getting rid of the dog.

Once he arrived back, Christiana let him know that he could be as mad as he wanted to be because his ploy had not forced his family out fast enough. She made sure that he understood that she did not give a damn about where he stayed or if he even returned. That is when she told him that he was no kind of a man for unhooking the stove to keep his children from eating and that he was a horrible person for unhooking the air conditioning and putting them in harm's way. Christiana reminded him that he knew it was a rough neighborhood on the south side of Chicago and that the killings were on the rise. After a couple of weeks of his cowardly abuse towards his family for whatever reasons, he finally found a piece of conscience and then hooked both the stove and air conditioning back up.

Di'Ban's dislike for Maximillia grew because she had become protective of the children and Christiana. He came home one afternoon after being gone for a few days. The children were in Cassius's bedroom, watching a movie along with Maximillia. Di'Ban came in from being out all night and then got mad that the house seemed peaceful without him. Meanwhile, Christiana was taking a shower, but she heard he had arrived at the house. Di'Ban stood at the doorway of Cassius's bedroom as he stared at Maximillia, trying to intimidate her. Maximillia sensed his threatening demeanor as she watched him closely, never taking her eyes off him.

Maximillia curled her top lip and her nose up, showing him her teeth. She began deep growling, warning Di'Ban that she was ready to defend herself and waiting for him to make any kind of a sudden move.

Di'Ban thought he wanted to challenge Maximillia, and once he acted like he was going to kick at her, she immediately charged at him. Cassius was sitting beside her when she rose for the defense mold, but he swiftly grabbed Maximillia before she got the opportunity to officially introduce herself. Niles hurried and closed the bedroom door because she was still determined to get Di'Ban. Di'Ban got angry that Maximillia never cowered down but met him for his challenge. After that, he left the house to get away. Christiana rushed out of the bathroom, asking her sons what had just happened because she had never heard Maximillia go off like that.

That was when the children said their father had come in, trying to intimidate Maximillia, but Maximillia won the intimidation challenge hands down while laughing. The boys started reenacting what took place, showing their mother what Di'Ban had tried to do but how Maximillia patiently waited for him to make his move.

Christiana came in that following Sunday afternoon from church and then started to prepare dinner as usual. She had begun frying the chicken that everyone loved, which they would come for each week. That was when the telephone rang. Once she answered the phone, the caller stated that it was Alley. She then demanded to speak to Di'Ban. Alley, however, must have remembered who she was trying to challenge because she quickly hung up before Christiana could say anything.

Di'Ban arrived home a few minutes later while Christiana was still frying the chicken. Without any hostility, she turned towards Di'Ban and said, "Alley just called here looking for you!" Di'Ban jumped, mad apparently. He did not appreciate her delivering the message to him. He reached over to cut the stove off to stop her from finishing cooking.

Christiana still had her hand on the handle of the frying pan with everything still in it. She said, "Trust me, but I'm finished frying this chicken!" Di'Ban believed that she was about to toss the hot grease and chicken on him. Christiana very calmly went on to say, "Di'Ban, I hope you realize that I'm just the messenger. However, I would advise you to get that cankerous whore in check!"

Immediately, Christiana went right back to what she was doing, preparing the dinner.

The frustration, due to years of mistreatment, was bringing out the not-so-good side of Christiana. She had grown tired of everyone's actions towards her. These times caused her to be done with people pleasing and living life for everyone else other than herself. Christiana realized that eagles fly solo and are equipped to handle themselves through any storms, regardless of the direction of the winds. That was how she equated herself.

Chapter 46: The Indifference

Christiana had some unhealthy thoughts about what she could do to Di'Ban and about how she should take him out of his misery. However, she'll find out that his misery will serve him cold by his own hands!

One night, while on a phone call with one of her friends, they were discussing how ugly his behavior was. She told India that she had thought about buying an industrial-size roll of saran wrap, hoping to catch him good and intoxicated, then she should saran wrap his ass to the bed! Christiana went on to say, "But I'm leave him a hole to be able to breathe out of, that is if he can!" India cracked up laughing but said, "Girl, your ass is going to jail!" Christiana said, I'll go sit on the curb and wait for the police to arrive, then I'll eat boogers and talk about all kinds of different topics in front of them. Bet I won't do much time, but he's not worth my freedom, as they both laughed!

Di'Ban came in from work on a Friday evening and saw that Christiana was dressed and ready to go out for the evening, and once she got into her car to leave that evening, he decided to follow her! Christiana saw that he was following her, but he only ended up at one of her girlfriends' home, not knowing that it was for a girl's night out. Once she got out of the car, she made sure to wave at him while shaking her head!

Later that same evening, Christiana arrived back at the house after she had gone out. She then found Di'Ban sitting in the backyard with one of his questionable friends named "Menace"! Menace was someone who just happened to appear one day when Di'Ban claimed him to be a close friend of his! Oddly, after many years of their relationship, she had never heard anything about "Menace" until then.

Di'Ban realized that she was back at the house, and he then immediately made his way to where she was in her bedroom. Christiana was about to get undressed when he entered her bedroom without knocking! Christiana said, "You need to knock before you enter this room!" Di'Ban instantly walked upon her with his nostrils flaring. That's when she pushed him out of her face and then went to pick up the telephone to call the police department. She could see what it was about to escalate into.

Di'Ban violently snatched the phone cord out of the wall, then shoved her onto the bed, but that was when Cassius opened his mother's bedroom door with a bat in hand to use as a weapon against his father! Cassius said, "Do not touch my mother." Di'Ban asked, "What are you about to do? Hit me with that bat?" Di'Ban did not know that Cassius had already made a call to 911, and the police were arriving right at that moment. Two police officers arrived; One talked with Di'Ban, and the other talked with Christiana!

The police who went to talk with Di'Ban saw that there were children inside. He listened to him, stating that it was his house even though they were married. One policeman came out to ask Christiana if she could find somewhere to stay for the night so that they would not have to disrupt the children.

Christiana agreed with the officer to not disrupt the children. She left and stayed at a hotel until it was check-out time the next afternoon. Once she arrived back at the house, Christiana saw where Di'Ban had ransacked her room from the night before, and he was out trying to figure out where she had stayed. Di'Ban arrived back shortly after Christiana had. He immediately came in looking to start another argument while the children sat in another room listening.

Christiana walked into the kitchen as he followed in behind her, talking "shit" while meaning to belittle her! Di'Ban said, "I would not touch your ass with a 10ft pole! Christiana swiftly responded and said, "Trust me "buddy" you don't even have that to worry about!"

That comment deflated Di'Ban's ego, so he tried to belittle her again by saying you are an unfit mother, but that was when things took a hot turn! Di'Ban was about to find out that one foolish comment was about to cost him his life because she knew what she had done to take care of the children and that he should have kept that comment to himself!

Christiana's temper went to a place that he had no idea that she had the capability of going to; her patience had always been thin, and usually, she was slow to anger! Christiana was much like her father Bash, but once provoked, slow anger was off the table!

The difference between Bash and his daughter was that she had endured much starting from birth, as she had watched him sit silent more than he should have! Christiana was learning to teach people how to treat her, even if it meant walking away and leaving them to themselves!

Christiana had begun breathing heavily as her vocals got low because she was trying to control her anger! Christiana slowly and with much conviction said, "Bitch you getting ready to die today, I promise you!" Christiana, without haste, said, "They are getting ready to pick your ass up off the front lawn in a body bag"! Immediately, she snatched the first thing she saw, which was a glass container from the kitchen counter. She cracked it on the wall and then charged after him in a rage!

Di'Ban's eyes were huge with fear. He realized that she was really going to attempt to take his life. She was on his behind, slicing at him while he was running, trying to get away! Di'Ban barely made it into the bedroom that he shared with Niles. It was only by the skin of his teeth that he got into the bedroom. However, he was not able to get the door closed because Christiana was pushing on it!

Christiana was trying to open it before he had the opportunity to lock it. She meant exactly what she said about taking him out of his misery and not giving up! Di'Ban was struggling to get to safety until Ishmael appeared; the youngest son intervened and became Di'Ban's saviour at that moment! Christiana heard Ishmael, "Saying don't hurt my daddy." That was the only thing that snapped her back to a calm mind state, hearing her baby boy pleading on his father's behalf!

Di'Ban heard once Christiana had walked away from the bedroom door, and once he thought the coast was clear, he came out of the bedroom foolishly, talking shit again! Di'Ban believed he was in a safe zone while attempting to redeem his manhood!

What he did not realize was that Christiana was only feet away, around the corner from where he was! Christiana was listening while he continued talking "shit"! Christiana eventually would come from around the corner and go after him just for talking "shit" because she felt if Ishmael had enough sense to try to save him! Christiana felt he should have had enough sense to have shut his mouth!

Di'Ban heard her once she made a sudden move. He then saw her come from around the corner! Di'Ban looked like a nervous deer in the wild during hunting season because he didn't realize that she was still that close! Once he heard her coming after him, it sounded as if she was running at the pace of a loose lion on a rampage after its prey!

Di'Ban felt that danger on the back of his neck as he hit the front door, running for his life! Christiana made it to the front door just as he had made his exit; however, he didn't even bother to look left or right as he ran out into the middle of traffic, trying to escape her! Christiana yelled out, "Oh bitch, you better run," while she still had a large piece of the glass in her hand!

Christiana stayed close to the front door, hoping he would try to reenter, but after that close call, he dared not attempt to go back inside, or at least not by himself! Di'Ban understood at that moment what she meant when she said that he was going to be leaving in a body bag!

Christiana heard him on his cell phone calling the police, pleading for help. He was telling them that she had cut him up with some glass and that he had her car blocked in! Christiana came back in from the front porch. She was surprised when she saw that her sons had already cleaned up the glass and then scattered it in the alley before the police had a chance to arrive.

Christiana called her father Bash. She told him, I am calling to let you know that the police are on their way here to the house, and they are probably going to lock me up because I just cut "Di'Ban's" up!

Christiana also said, "If I don't call you back, could you please come to get my children!" By this time, the children were already seated on the couch. She then took a seat alongside them because she saw that the police were on their way inside the house!

Chapter 47: No Possession of Heaven or Hell

Christiana saw an older black policeman and a younger white policewoman, and when they walked in, the policewoman instantly unsnapped her holster as if Christiana was a threat. She then immediately began talking crazy to Christiana.

Christiana calmly said, "First off, you need to bring your tone down when speaking to me, and the unstrapping of your holster was very much uncalled for, seeing that I'm sitting here with my children!" The policewoman kept asking Di'Ban questions in his favour because Christiana did not bow down to her while her partner stood still, watching and listening.

The older policeman was paying attention to everything that he saw but quickly identified that Di'Ban was a narcissist trying to manipulate the situation! The older policeman immediately took control of the situation, seeing that his partner, along with Di'Ban, was attempting to make things worse for Christiana in front of the children.

The older policeman then told Di'Ban, "You are going to get all your vehicles out of the driveway, and you're going to get enough clothes for a couple of weeks, and do not come back, or you will be arrested."

Di'Ban could not believe that he was unable to get Christiana arrested! All he kept asking while sounding like a disturbed lunatic was, "Where's the glass? Where's the glass?" No one answered him like they had no clue as to what he was even babbling about.

The male officer told Christiana that she needed to go and get a protective order first thing the following morning because he would hate to return there because things got out of hand, and she ended up killing Di'Ban!

Di'Ban left for a short time but returned shortly after the police had left. He then went over to an older female neighbor, Ms. Hattie's house, who lived across the street. Ms. Hattie was a woman who did not care for Christiana for unknown reasons other than for whatever he had told "Ms. Hattie" about her! Di'Ban was over on her front porch, letting her wrap his wounds while they both were talking loudly about Christiana, trying to get a reaction out of her.

Shortly after the altercation, another neighbor, Shantell, was paying attention to what was going on. She called Christiana to check on her and to tell her what she heard them on the porch saying about her. Christiana told her, "It's ok, Shantell, I'm not coming out because if I do, all it would take is one good sarcastic remark from either of them for me to run across that street and cut up everybody's asses, including Ms. Hattie with her old bearded gray ass!"

Natalie immediately jumped on the phone after Bash told her what was going on with Christiana. She then called Nadia to tell her that Christiana was about to get locked up! Bash or Natalie never called back to check if Christiana had gotten locked up or if their grandsons were out of trouble.

Nadia picked up her phone and then called Christiana, laughing, saying that all that she wanted to know was if it was like Okay on the Corral, referencing a cowboy brawl. Nadia certainly made sure her sister understood that she was not calling to check on her or find out if any of her nephews were okay, either. Christiana said, "As for right now, Nadia, I don't find any part of it funny. Therefore, I'll let you go."

Apparently, Bash and Natalie felt that she was the butt of all jokes. They then jumped on the phone to call other family members who didn't even live in the state of Illinois just to run down Christiana's relationship drama!

Christiana received a letter about a week later from her uncle Jacques Bouvier Jr. Esq. about the altercation with all the details that he was told about! Jacques Jr. Esq jokingly stated that he would like to be her sponsor for her fights, even with giving her a stage name!

Christiana was sick and tired of being the butt of everyone's jokes. However, she knew that she was not the only one in the family with some sort of problems. She just didn't speak on their trials as a joke. Christiana also understood that everyone gets a turn and that it might be her for now, but theirs would come out of hiding soon enough!

Di'Ban received his divorce documents when Christiana filed them; however, he and his attorney contested them, which meant that it would drag out for a while. During that year, Di'Ban continued with his hateful ways! Christiana understood that both clocks were ticking and hers had already rung!

Christiana was looking for a new place for her and the children to live and for new employment, but this time, she was also looking for out-of-state employment. This time, she wanted a new start for a different chapter of her life. She always felt that she was destined to be somewhere else that would bring joy and purpose to her life.

Christiana got an interview for a company that was looking to fill positions in Seattle. She was excited about the possibility of having a change of scenery! She wanted to expose her children to larger things in life and show them that they should never stay in one place because it is a big world that they live in!

Christiana talked to her kids about many things in life often, hoping to broaden their minds. She lived her life with them as an open book so they would not make the same mistakes as she had! Christiana hoped that they understood that it is not about how much one can put their hands on but that it is rather what is within you that is the most important.

Chapter 48: Light and Dark Can't Come Together

Di'Ban had been gone for an entire weekend with 'Alley,' once he arrived home, he saw that 'Christiana' was up and dressed professionally! Christiana had an interview for a large company that was looking for agents to work at their Seattle office; she was also getting 'Ishmael' ready for daycare so that she could be on time, and everything was running smoothly until Di'Ban returned.

Di'Ban came in with a hostile tone. Once he saw her dressed, he asked, "Where are you going?" Christiana completely ignored him as if he was not even speaking to her. Even after asking her for a third time about where she was going, she still would not answer!

Di'Ban went and picked up the telephone to call the police; he started telling them that he had someone in his house that he did not want there. The police showed up, and that is when Christiana, along with Ishmael, calmly stepped outside to talk with one of the police officers because she already knew the drill.

One police officer asked her, "What's the problem?" And that is when she stated that the problem was in the house. The other police officer who had gone inside to see what was going on that caused Di'Ban to call them. He came back out looking shocked by what Di'Ban had said to him!

The police officer said, "Ma'am, he wants you to leave. Today!" Christiana said that's fine, but I will not leave without my children or our belongings! The police officer then asked Christiana to step back into the house with him. The police officer told Di'Ban she would leave, but you'll need to give her time to get the kids and their belongings!

Di'Ban played this calm, collective role in front of the police officers, looking like the corners of his mouth were curled up! Di'Ban told the police that it was not a problem to allow her the time she needed to get the belongings right before the police left.

Christiana unfortunately missed the opportunity for the interview that she was scheduled to have. She immediately went into her room to change her clothing; however, she remained quiet while not saying anything to him because she felt a change was coming.

Once again, he started talking ignorantly to her because he could not get a rouse out of her. He kept following in behind her from room to room, saying you need to hurry it up and get this shit out of my house! Christiana continued to ignore him while moving to gather their belongings.

Christiana first began packing the kids' belongings while she was bringing their dresser drawers out into the living room, and he would still not leave her alone as he stayed focused on tormenting her. Di'Ban's last-ditch effort to get a rouse out of her was when he said, "I'll throw all this shit out on the lawn!" Christiana stopped in her tracks; she looked at him and said without hesitation, "Do not put your hands on none of my kids, shit!"

Di'Ban said in a threatening tone, "What are you going to do about it?" Again Christiana said, "Like I said, don't put your hands on none of my kids' shit!" Di'Ban reached and grabbed two of the kids' dresser drawers to throw out onto the yard. Christiana took in a deep breath then said, "Mutha fuc.." as she ran upon him with a closed fist punching him directly in the face with all her might, then finished her comment with your hands on none of their shit! Di'Ban again ran to the phone to call the police; he was sure to get her arrested while telling the operator that he had "Tyson" in his house and that she just attacked him!

Christiana returned to packing their belongings, and once more, the same police officers who originally came returned and asked Christiana what was going on. Christiana felt everyone was against her because they kept asking her what the problem was!

This was not a go-along-to-get-along moment for Christiana, and she said, "Listen, I'm not afraid of no one in this room. Now he asked for us to be gone, and I'm in here trying to get our things together to leave. Now, what does not make sense to me is that you all can see that's exactly what I am attempting to do. However, he is in here trying to intimidate me by throwing my children's belongings out on the lawn, as you both can see."

The police decided that they needed to take a different measure and took Di'Ban outside to leave her alone so that she could get her belongings in peace. They also told him that if they had to come back and if he didn't leave her alone, they would lock him up.

Time was ticking, and school had just let out. That was when Christiana looked out and saw that the two older children's school bus was pulling up! The boys were looking at what was going on, and so was everyone else on that bus. Cassius and Niles both told their mother how embarrassing it was for them and that the other children were asking them why their stuff was out on the lawn.

Christiana could not get any help moving on such short notice, nor could she get a rental truck on short notice either; she then called and asked her father 'Bash' if she could use his truck, explaining that Di'Ban was putting her and the children out! Christiana also wondered if she and the children could stay with them until she could find a place. Bash told her she could use his truck and that they could stay there with them.

Christiana took Ishmael to leave him with her parents, leaving the two older boys at their fathers to help pack their belongings. Christiana put some items inside her car to take with her before dropping 'Ishmael' off. Once they arrived, her mother 'Natalie' instantly started with her foolishness to further exacerbate Christiana's stress. Christiana told her mother that she did not need her stressing her out any further than what she was already enduring and that if she did not want them there, she should just say so and she would look for other alternatives.

Bash spoke up; he told his daughter to go ahead and finish getting their things. He said he would make room to store most of their things inside the garage. Natalie was still creating chaos for each trip that Christiana made; she kept asking her daughter how much stuff she was bringing. Christiana said, "I have no idea. I'm trying to get all our belongings so I don't have to go back!" Christiana was making multiple trips with her son's help, and after returning at one point, Di'Ban had some of his friends over with him, sitting in the living room having drinks like it was a party, even though they could see what was going on. Christiana walked back in to get more items, and once she noticed what was going on, she said out loud, "***Misery loves company!***"

Christiana had gone into the kitchen to load up some items, mostly that they had received as wedding gifts, but she was handing those items to one of the children's friends to take to his mother because she did not want any reminders of that day.

Christiana was at the counter, packing while smoking a cigarette. Di'Ban decided he would come and start some more drama now that he had an audience! Finally, he walked into the kitchen and said, "Don't be smoking in my house." But Christiana just kept packing while ignoring what he was saying.

That's when Di'Ban decided to walk up and smack the cigarette out of her mouth, but that was a huge mistake on his part! The children stood there in the kitchen watching it happening, looking shocked!

Christiana didn't say anything when he smacked the cigarette out of her mouth, and she immediately snatched up a heavy crystal container, then swiftly cranked it back about to serve him a pitcher's throw to the head! Di'Ban saw what was about to happen to him; he then made the matrix move because he realized that she was about to knock him out, which would probably cause him some severe injuries. Christiana immediately and calmly set it back down without missing a beat. What he didn't realize was that it was only because the "Most High" told her to sit down!

At that point of his foolish choice to assault her Di'Ban graciously and expeditiously strutted his trouble-making ass out of the kitchen after that! The boys and their friend went into

Cassius's room mocking their father's matrix move along with the scared face as a bonus.

Christiana was almost done collecting everything. She needed to get Maximillia and her ten puppies, who were only a few weeks old. She told the boys to make sure to get Maximillia's cage. Di'Ban asked the boys what they were going to do with the dog's pallet on which she had given birth to the puppies. That was when Christiana said, "We're leaving it here for your nasty behind to sleep on."

Once Christiana arrived back at the house, she gave her parents seven hundred dollars to allow her and the children to stay there with them. The following week, Christiana had to rise very early in the mornings to drive the children from the far north side of Chicago to the far south side of Chicago to get them to school on time in the deep winter with single-digit temperatures.

Christiana would return to the house after dropping off the boys and then make sure to feed Ishmael. She would not even have time to feed herself or to get dressed because Natalie was adamant in creating even more stress on Christiana, complaining that she needed to go outside to clean up frozen dog crap that was at the very back of the yard near the alley where the dog would do her business. Christiana did not have a problem with cleaning up behind the dog, but she would at least wait until she had eaten and prepare herself for the outside temperatures. She did it once asked to try to keep her mother from complaining.

As soon as Christiana went out cleaning up behind the dog, Natalie would come to the door telling her that she needed to go and bathe Ishmael to get him dressed; it was constantly one thing after another with her mother.

As soon as Christiana finished bathing herself, she would go into the kitchen to prepare dinner for the other children to arrive home to eat! Christiana, daily, would go to do their laundry to keep order in the house before going to retrieve her sons from school. She was having difficulty figuring out how she would be able to schedule interviews because of her schedule.

Chapter 49: Home Team

One evening, Christiana's neighborhood friend 'India' called her, stating that her mother was sick and was dying! Christiana told India that she would sit with her after she picked the boys up and fed them. India lived on the same block as Christiana's parent's, and they knew her mother.

Christiana had gotten the children situated after dinner and in their pajamas to watch a movie and to keep them occupied! Christiana asked the boys to keep an eye on Ishmael until she arrived back at the house.

Christiana told her parents that she was going to walk down the street to visit India for a little while because her mother was in the hospital, and she was not expected to survive throughout the night.

Thirty minutes later, Natalie sent Cassius to go and tell their mother that she needed to come back to the house. Christiana returned to the house to find out what was going on that she needed to return so suddenly. Natalie said because these kids are here. She then asked Christiana how much longer it would be regarding India's mother.

Christiana, unbelievably, looked at her mother. She said, "Her mother is dying as we speak, so I don't know how much longer it is going to take!" Christiana told Natalie, "The boys were OK, and they would be falling asleep soon. Besides that, they know what to do to take care of themselves until I return!"

Christiana said, "I am preparing to take India to the hospital. I will most likely be there with her until however long it will take. I would hope that someone would be with me at a time like this."

It was not going to be long after India and Christiana arrived at the hospital that Christiana would see that it was going to be at any moment that her mother was going to transition into death. Christiana was honored that India wanted her at her side at such an unimaginable time! Christiana saw the moment once it was at hand that her mother's heart rate was depleting rapidly and about to stop. Christiana immediately told India you need to hold your mother because she's about to transition!

Once Christiana arrived home after dropping off India a little after midnight, Natalie asked, "How did it go?" Truthfully, Natalie was only interested in knowing if India had fallen apart upon her mother's death! Christiana realized just how insensitive her mother was; she also realized that everyone gets a turn, and that's when they will understand what a person goes through losing a loved one.

Chapter 50: Misery's Company

Christiana was in the kitchen cooking to take a dish down the street for India and her guest when her cell phone rang. She answered, and to her surprise, it was Di'Ban asking if they could get together to talk.

Natalie was sitting at the dining room table, eavesdropping as if it was a secret, but Christiana never tried to cover up who she was speaking with! Christiana told him that if he wanted to talk with her, he could come the following day to church. He agreed to meet her there at church.

The next morning, when Christiana had gotten up to prepare for church service, her father Bash came to her bedroom door; he told her he needed to talk to her! Christiana already knew it would be some nonsense regarding Di'Ban, but to the extent it was about to go to, it would come as a surprise!

Christiana sat down to listen to her father; he said, "I know that Di'Ban called here yesterday." But that was when Christiana said it was no secret! Christiana's father said, "I'm going to tell you something. Di'Ban came to me one day and said the only reason that he hadn't married you sooner was because he said that you were a slut!" Christiana looked at her father, heartbroken at that moment as she lost respect for him!

Christiana understood very well that in the same way you can tell when a child is lying to you, you can also tell when an adult is lying to you! Christiana never thought that her father would ever lie to her. One thing that she understood was that no man would ever allow some other man to speak such foulness against their daughter directly to his face and then leave him with teeth in his mouth, but if that was the case, then he had completely failed her as a father!

Christiana felt deeply hurt by what her father had said to her, and when she tried to talk to Nadia about how their father hurt her, Nadia climbed upon her high horse, curled her lips up, and said, "I don't know nothing about that, but I know that I would never deal with the things that you have dealt with!" Christiana stopped trying to talk to her about how hurt she felt because she realized her sister had turned a deaf ear and showed no empathy in trying to understand what she was saying.

Di'Ban showed up at the church, and after service, he wanted to take Christiana out to eat and to talk! Once they reached the restaurant and ordered their meals, Di'Ban claimed that he knew he'd been a horrible person, but if she gave him another chance, he would show her that he could be the man she needed him to be. She understood that he just needed to be a better man for himself and for his children!

He also stated that if it takes more counseling, he was willing to do whatever it takes to make the marriage work. Christiana agreed to give it one last try, but that time, she also listened to what the "Most High" had set in her spirit, which was that he would show her Di'Ban in a 3month period; however, she did not withdraw the divorce or the child support because it was already into place!

Christiana did not mention to him what she had been told by the "Most High" regarding him being shown to her in a 3-month period! She then waited and watched him as he transformed back into the same man he claimed he had been changed from, revealing himself daily from the start.

After a few days, Christiana told Di'Ban that she needed him to keep the older boys throughout the week to get them back and forth from school because it was not allowing her to schedule interviews, as her day consisted of rigorous ripping and running. This was the first and last time he ever had to take on the role she had always managed by herself.

Christiana said she would make sure to pick up their laundry on Friday evenings to get their clothes ready for the following week. In the first week, the older boys were left with their father. Christiana went to pick up the laundry at the end of the week as promised because she wanted to ensure that their clothing was neatly ironed and presentable for school.

Christiana arrived back at her parent's home with her youngest son, Ishmael. She then put him to bed and began the other boy's laundry so that she could return it to them the following day. Christiana was in the laundry room when her father, Bash, got out of bed and came into the laundry room to tell her that she was no longer allowed to do her children's clothing there.

Christiana thought about the fact that she had just given her parents seven hundred dollars, and it had only been a couple of weeks since they had moved in. She could not believe that her parents could be inconsiderate about allowing her to do the children's laundry there; both of her parents knew that she was the type of mother who made sure her children's needs were met, and she did not want to have her son's worrying about clean clothes for school.

Christiana also knew that it was none other than her mother who orchestrated the plot because she was the one who always got Bash to do her dirty work. After all, she was the one who always threw the rocks and then would hide her hands.

Christiana said to her father, "Well, if I'm not welcomed to do my children's laundry here, I probably need to leave because this does not make any sense!" She then asked, "What will it be next? I guess next my kids won't be able to eat here either!" she disgustingly said, "I may as well have stayed in the hell hole that I was already in because I can't find any peace here with my parent's!"

The following day, Christiana also received a call from one of her good friends 'Naomi' who said, "I tried to call you at your parent's house, but your mother told me that I could call and catch you at Di'Ban's house. I was so confused because when I called his house, he said you no longer lived there! Di'Ban told me to try your new cell phone and then gave me your number." Naomi continued, "Your mother stated she didn't know what was wrong with you and that you had already moved back with Di'Ban."

Christiana told Naomi that's a lie because I am here with them right now! Naomi said, "I don't get why your mother would say such a thing, and I don't have any reason to lie on her!" Christiana said, "Naomi, I believe you, and I am going to address this because she has been doing nothing but creating another hell for me since I've been here."

Christiana went to talk with her parents and asked why her mother would tell someone that she had moved back over to Di'Ban's. Natalie immediately jumped on the defense, stating that she didn't tell anyone anything and that people needed to stop lying on her.

Christiana also questioned who had called, putting her business out to the family that doesn't even live in this state because she herself hadn't called anyone. Again, Christiana was told more blatant lies, stating that they hadn't talked with anyone regarding her business. What they did not know was that she had received a letter from her Uncle Jacques Bouvier Jr. Esq.!

Christiana realized that she had just traded one hell for another; she figured that she should move back to Di'Bans so that it would at least alleviate the stress of having to spend her days running from one side of town to the next all the while getting her children clean clothing to wear and also making sure that they were fed. Christiana was going to put Ishmael back into daycare in a hurry to find and gain employment as part of the process; she had become fed up with everyone, with the yoyo rides, and understanding that she was prolonging her own growth because of it.

Di'Ban noticed Christiana's evolving strength and how silent she had become. He asked if they could seek counseling to work out their problems, and she told him that if they did that, it would be with her minister. Christiana knew that her minister had always talked about how he had done and seen it all in, that not much would get past the minister, and he could lead her in the right direction!

Christiana talked with the minister and then made an appointment for them to speak with him. Di'Ban got in front of the minister and thought that he could bullshit him about all that he claimed he did for her!

The minister sat there, letting Di'Ban believe that he was fooling him, while he kept watch of Christiana's body language, seeing that it was sickening her! That was when the minister said, "Alright, Di'Ban, I've heard what you've had to say, but I really would like to hear what Christiana has to say right now!"

Christiana looked directly into the eyes of the minister and said, "Listen, the "Most High" gives us all wisdom and knowledge, but he also allows us to see the devil coming, which he comes pretty and dressed up!" The minister's eyes widened as he put his hand to his chin, visibly showing that Di'Ban was unprepared for her response. The minister sat back in his seat, knowing that she was completely done with all the circus acts!

Chapter 51: The Camouflaged

Christiana had taken ill shortly after moving back into the house with Di'Ban. She had to be hospitalized for a week due to one of her major organs failing. Once she was released from the hospital, she went to pick up a prescription and took Ishmael with her.

Upon returning to the house, Christiana noticed that she could not pull into the driveway because 'Menace' had pulled into the driveway. Menace was standing in the front yard as Christiana and Ishmael pulled up. She sat there for a minute, thinking that he would see them, and then move out of the driveway so she could park her car.

Once Christiana saw that he was not attempting to move his car, he ignored her as if she was not sitting there! Christiana let her car window down and politely asked Menace if he could please move his vehicle to let her in the driveway.

Menace ignored her asking and kept smoking his cigarette. Christiana said to herself, "This bitch done lost him damn mind!" Christiana put her car into park gear and then got out of her car. She entered the house where Di'Ban was and said, "Listen, I don't know what your '**friend's**' problem is, but I asked him to move out of the driveway, which he's apparently refusing to do."

Di'Ban looked at her and said I don't have time for this bullshit! Christiana waited for a second to see if he would tell his '**friend**' to move out of her way, but he continued with what he was doing from when she first walked in!

That is when Christiana said, “You know what, don’t even worry about it, I’ll take care of the shit myself!” Christiana walked back outside then said, “Listen, Menace, I am the woman of this house, and I’m pretty certain that you would not allow Di’Ban to come to your house to disrespect your woman or wife, so I would appreciate it if you move out of the driveway so I can park my car.”

Menace said, “I am not Di’Ban, I will fuck you up!” Christiana said, “Bitch, you are going to fuck up who?” They both went toe to toe in a heated argument at that moment, and that was when Di’Ban finally appeared outside, running and jumping between Menace and Christiana, saying, “Don’t yawl do this!” Christiana said, “What the hell do you mean, ‘Don’t yawl do this’?” Di’Ban told Menace, “Man, let’s go!” Di’Ban then got into the car together with Menace, and they drove off!

Christiana was thinking, how do you allow another man to disrespect your wife, then you get in the car with him and leave? Christiana thought, now that is suspect; I guess I just saw where his allegiance lay, and she understood that it was not with her!

The neighbor Victor who had given Christiana his dog Maximillia walked down to her and said, “Them bitches are gay for that shit!” Victor said, “My boys and I were waiting on his ass to swing because we were going to beat the shit out of him and Di’Ban if need be!”

Victor asked where Maximillia was. Christiana said, "I was not even thinking of bringing her out with me, but I wish I had because she would have lit his behind up." One more neighbor, Linda came across the street and said, "Christiana, I never knew that you could raise your voice that loud, but Di'Ban was way out of order to allow another man to talk to you like he did. However, I saw those young guys were ready to light both their asses up."

A week later, the boys and their mother were riding down the street. Right as she was slowing down to approach the driveway, she eventually came to a complete stop! Christiana and her sons had looked down the block and saw something that seemed very odd, which caused Christiana to question Di'Ban's sexuality by what caught her attention, which made her stop!

Christiana noticed Di'Ban and Menace slowly strolling, sashaying together while licking ice cream cones. Christiana sat there in shock for a minute. She then said out loud, "Oh... that's gay," but it's all making sense now. I was just a smoke screen for him!

Chapter 52: Freed from Bondage

The day arrived when Christiana was granted the divorce after close to a year; the day that the decree came, she was pleased and didn't feel anything or any way in regards to it because she had been done, got off of that ride a while ago however, Di'Ban had a different emotional response!

Once Christiana arrived home from work, she entered her bedroom, and then Di'Ban came and knocked on the bedroom door to speak with her. Christiana opened the bedroom door to see what he wanted. She noticed the decree in his hand. However, she knew the reality would impact him because he thought he had her underneath his thumb because of the children.

Di'Ban began pleading his case and wanting to reminisce on a watered-down version of what he claimed to have loved about her! Di'Ban realized that his words had finally fallen on deaf ears because he had seen but felt that his time for *any more time* with her had expired!

Di'Ban witnessed firsthand the lack of interest in anything coming from him, and she looked at him with the most annoying, disgusted demeanor. Christiana's only response was, are you done which fucked Di'Ban up once he had met the new Christiana!

Di'Ban didn't understand that Christiana had finally realized her worth! Christiana was sickened when she realized she had gotten down in the dirt with a dog!

Christiana always knew that he was out of his league with her, but she also understood that he never told her he was the man she thought he should be! Christiana took responsibility for the role that she played in allowing him to mistreat her and for what she considered to be a trifling hemorrhoid of a baboon's ass for treating not just her but more so for how he treated his own children the way that he had.

Christiana first apologized to herself and then refused to waste any more time with anyone who couldn't ever meet her at her level. She also understood the meaning of being unequally yoked was broad! Christiana hoped that her son would not follow in his footsteps, but she would learn some things about him much later.

Christiana knew she needed but wanted better for her life! The one thing she did not want was for her sons to look at her like she was stupid and for them to think that it was OK to treat a woman and their children in this same manner. Christiana explained to her sons that if they ever treated their wives and children in that manner, she would be done with them because they had seen what it was like.

Chapter 53: The Reality

Christiana was at work one day when she received a call from one of her neighbors, who happened to be a relative of one of Christiana's neighborhood friends, Donika! The neighbor told Christiana. "My cousin Donika needs to speak to you." Christiana immediately guessed what it was possibly about.

Christiana called Donika to see what was going on, and that is when Donika said you know that you and Kyrie were supposed to be married! Christiana laughed and then said, hey, that was on him! Donika said I had gone out of town, and I happened to run into Kyrie, and before he could ask anything about me, he was asking about you! Donika said he is coming to Chicago in a couple of weeks for a weenie roast that I am having, and he wants you to come because he really wants to see you.

Christiana asked Donika if Kyrie was still married because she didn't want any problems with an upset wife if she showed up. Donika said, I don't know about all of that, but are you coming?

Christiana said, I'll come; it will be nice to see Kyrie, too!

On the day of the weenie roast, Christiana was nervous about showing up because she did not know what to expect. She stopped on her way to Donika's event and purchased herself a couple of shots, trying to calm her nerves! Christiana knew she did not drink often, but she desperately wanted to calm down!

Once Christiana walked up into the yard, those shots kicked in on her behind! It felt like the stronger the buzz got, the more she saw more of his family members were present! Kyrie was at the picnic table fixing himself a plate; he had not seen that she had arrived yet! Christiana said to herself let me just go ahead and get this over with so I can go sit my ass down, hoping that the alcohol would wear off!

Christiana walked over to the table feeling like she couldn't almost breathe. She knew that all eyes were probably on her, and that's when she tapped Kyrie on his shoulder and said hey! Kyrie turned to look, and his face lit up once he saw it was Christiana!

Kyrie put down the plate that he was preparing and literally scooped Christiana up into him, holding on to her with an extended hug! Christiana didn't know if she should embrace him back or not because she didn't know what the deal was yet with him.

Christiana thought well, this is the perfect time for me to find out, she asked Kyrie where is your wife while you are holding on to me like this? That was when he said, "She's not here." Christiana thought, here he goes again!

Kyrie said, please come sit with me; he did not want to let her get away from him, at least for the evening! While she was following him, those "two shots" kicked in with full effect on her behind, leaving her feeling lightheaded!

They took a seat side by side, but as soon as Christiana sat down, she saw Kyrie's mother! Right before she got the opportunity to speak to her, Kyrie said, "Mama, do you know who this is?" Kyrie's mother said, "I know exactly who that is, that is that little pretty 'Christiana Bouvier'!"

Kyrie's mother was talking to her briefly before Kyrie apparently smelled alcohol coming from Christiana, and he asked her if she had been drinking. Christiana said, "Yep! I was nervous, and I think I drank a little too much." Because of her comment, Kyrie laughed and said, "It's OK, just let me finish eating this plate, then we can go for a walk!" As soon as he was done eating, he took Christiana by her hand and then led her away from everyone.

Kyrie wanted to spend some time alone with Christiana, and once they made their way out of the yard onto the smooth pavement, he hooked his arm into hers as they began doing what they had always done best: walking and talking with peeled-down layers towards each other!

They came to a corner that had a brick wall which they could sit on. They took a seat and began reminiscing, which was evident that their hearts remained entwined. Kyrie and Christiana hugged as they embraced with no measure of time for the event!

Christiana told Kyrie that she needed to use the restroom. When they walked back into the house, Donika, his mother, and Auntie were in the kitchen with bright lights on. Donika said out loud, "Damn, Christiana and Kyrie, where have you two been? Did you try to eat his ass up or what?" Laughter rang in the room!

Christiana was unsure of what they were even talking about until she turned and looked at Kyrie. She saw that he had lipstick prints all over him! Kyrie just smiled, and Christiana felt embarrassed, then went straight to the restroom. And once she came out of the restroom, Kyrie still wanted to spend more quality time with her.

They went to Christiana's car, listening to music while they finished talking. Kyrie then leaned over and began kissing Christiana, but as soon as she started to enjoy it, Christiana realized that she would not be a sidepiece for him or anyone! Christiana pulled away, apologizing. Kyrie said, "Please don't apologize because I didn't do anything that I didn't already want to do!"

The next day, Christiana felt down because she knew that she was what is known as the one who got away! Two weeks later, Natalie called Christiana hatefully, saying, "Kyrie came over here looking for you. He left a number for you to call him." Christiana was shocked to hear that he had returned so soon!

Christiana called to talk with Kyrie, and he stated that he wanted to see her again but he would only be in town for a couple of more days. By that time, Christiana did not need a drink to calm down. However, she also wanted to see him, and she refused to get caught up in their nostalgic feelings.

Kyrie was at one of his sister's houses, where he stayed, and once she visited him, he received a personal phone call. Christiana couldn't help but hear, and once she heard Kyrie say that he loved whoever it was on the other end of the call before he hung up the phone, Christiana decided to leave!

Christiana didn't expect Kyrie to walk her to the car, but he decided he wanted to, then got into Christiana's car with her. They talked for a little longer before Kyrie leaned in, wanting to kiss Christiana, but she only offered him her cheek that time!

Christiana knew that Kyrie wanted to feel the magic of their kisses once more, but she was not about to let him mess with her emotions! Christiana was sure that he couldn't understand how she did not want to kiss him, but she also knew that he was denying himself the truth that she was the love of his life!

Kyrie told her on a phone call much later that they didn't make it because she would not have been able to handle it, and Christiana, without question, just let him believe the lie that he was telling himself!

Chapter 54: Seasonal Ties

Christiana was the one who always had to take the boys to the barber shop, while their father always only took himself to get his hair cut. Christiana had started sending money with Cassius, the oldest of her sons, to pay for all three of their haircuts because the men were always flirting with her, making her uncomfortable.

During one shop visit, Christiana forgot to give money to Cassius, so she had to go to the shop once she arrived to pick up her sons! When she walked into the shop, all the barbers were happy to see her and greeted her, saying, we thought we all had done something to you!

The boy's barber, Aaron, personally knew her from where they had grown up together. He wanted to know why she no longer came into the shop. Christiana told him that it is a place for men and that she feels out of place.

After that, Aaron asked her what she had been doing, and she told him she was looking for better employment and looking forward to moving into a new home since her divorce was final! Christiana also told him that she was waiting on her test results from a major corporation in the Chicago area.

That was when Aaron said, "My client Tony, right here. He works for that corporation also!" Aaron called for Tony to come to meet Christiana. He then let Tony know that she's waiting to hear back about employment with the company that he works for.

Once Tony came to speak with her, all the other men took notice. That was when he asked Christiana if she wanted to talk about the company's employment process.

Tony only asked because he could tell she was uncomfortable with everyone listening to their conversation about how the company handles applicants and the process for once you gain employment. Through their discussion, they found out that they both knew some of the same people; he then offered her his phone number if she had any other questions that he could possibly help her with.

Their conversation was pleasant, and Christiana enjoyed talking with Tony. She had an event that she would attend later that evening, and she thought she would like to see if Tony would like to join her as a guest.

Christiana called him; however, she was not able to reach him. Tony returned the call the following day, stating he hated that he missed her call! Tony said he would have gone with her because he enjoyed their conversation at the shop; before they knew it, they had talked for about an hour about an array of things.

They even talked about the fact that she had just gotten out of an abusive marriage and that she was not looking for anything, and he told her that he believes that if something is meant, it will happen just to stay focused!

Christiana explained to him how she hated that she had wasted most of her youth with an ex and how she had listened to the opinions of others instead of listening to herself! Last but most importantly, they talked about the "Most High"!

Tony asked her if she would like to hang out with him for a little while once Friday arrived. She told him she could, but it would not be until later that afternoon because she was getting ready to travel out of state the following day. Christiana told him she needed to take her kids to a sitter first.

Christiana did exactly as she had stated. On Friday, she called Tony, and they decided to get something to eat and then rode around viewing Lake Shore, but with nothing else in mind, they enjoyed looking at some homes together and talking about each other's likes and dislikes.

Later that evening, they decided to take in a movie while enjoying each other's company; once they were done at the movie theatre, Christiana told Tony she needed to be headed home because she would be leaving to travel out of state in a few hours.

Tony asked if she could call him once she arrived at where she was traveling to, just to ensure she had made it safely.

Christiana had arrived back at the house, and she had already packed for her trip earlier in the day and had her luggage sitting in the living room, ready to go. Once she returned from her outing, she changed into some clothes to be comfortable for her ride.

Di'Ban was home lying on the living room couch; meanwhile, he was fuming because he saw that Christiana was living her life without any interest or concern for him!

Christiana came into the living room and then took a seat, but she quickly noticed how agitated his body language was, with his nostrils fairing. Christiana was watching the television when he decided to ask her with hostility, "Where have you been?"

Christiana looked at him as calmly as she knew how; she said let's not start that because I don't ask you anything about where you've been. Di'Ban decided he would throw a threat at her and then said, "Yeah, when you get back…" but before he could finish, Christiana said, "I'll just be gone!" Di'Ban could not handle the fact that he no longer had any control, and she was letting him know it.

Di'Ban jumped up from the couch, picked up her coat, and threw it at her, but Christiana caught it. He then picked up her garment bag and threw it at her, but again, she caught it! Di'Ban then picked up her heavy cosmetic case and got ready to throw it at her. However, she knew that it would hurt her, so she jumped out of the way of him throwing it and then punched him in the jaw before he hurt her.

The fight was on between the two of them, but he still hadn't realized that when a woman gets into a fight with a man, she is fighting for her life! Di'Ban was having a hard time controlling the fight that he had started, so he grabbed and slammed her into the couch, then began digging into her arms with his nails.

Christiana kneed him in the growing area to get him off her, and she went into her bedroom to look at her arms in a mirror; she saw where she was bleeding from where he had dug into her skin with his nails! While looking to see her scars, she was saying to him let me have any scars, and believe me, your ass is about to go to jail.

Once she saw the scars, Christiana picked up the phone and called Chicago's finest to the house. The police arrived and looked at her arms. They then went and talked with him.

Meanwhile (Christiana) was telling them that she wanted to "press charges" for the abuse. One police officer gave her a card and told her to call on Monday to talk with the prosecutor's office.

Christiana grabbed as many of her belongings as she could and then loaded them into her car. She called Tony to have someone to talk to, telling him what she had just gone through as soon as she arrived back at the house.

Tony asked Christiana if she was still leaving to travel and that he wished she would stay, telling her that she could come to his house, but Christiana told him that she needed to keep her prior commitment!

Sunday night, after Christiana was about to return to the city, she knew for sure that she was never going back to Di'Ban's house other than to get her children's belongings. Christiana understood that sacrificing herself and trying to keep a family together did more harm than good.

Christiana called her childhood friend Sicarri to ask if she could come to stay there with him until she could get on her feet. Sicarri opened his door to her, and once he saw what Di'Ban had done to her, Sicarri said that it was abuse and she needed to be done with him. He also said that she and the kids are welcome to stay there until she can find housing.

Sicarri worked a shift from 2 a.m. until 2 p.m., and they rarely saw one another! Christiana retrieved her sons from the sitter the following day and then took them to their father's house. Christiana explained to her sons that she was not coming back over there, but she was working on getting them a place. She took Maximillia with her to Sicarri's, and Maximillia stayed in the bedroom with her.

Di'Ban had lost all control of his mind; he had begun riding around to everyone that he knew Christiana would possibly go to. Christiana and Maximillia were in the room asleep when Maximillia's metal collar made a jingling sound that woke Christiana.

Christiana watched the dog to see what grabbed her attention. When she did not hear the garage door open, she realized that it was not Sicarri who was arriving back home. It was around midnight, and he would not come back until much later, so she got up from her pallet from the floor, and Maximillia followed her as the dog was still on guard.

Di'Ban had gone, then jumped over the fence into the backyard and climbed upon Sicarri's deck; he was banging on his patio door like a madman! Christiana still did not know it was him; she stood still without cutting any lights on to see what would happen next.

Di'Ban got no response, so he made his way to the front door, banging on it. That's when Christiana asked who was at the door, and he said it was Di'Ban! Christiana and Maximillia were at the top of a short flight of stairs, and she went down to the door while the dog stayed put for a minute to find out what he wanted.

Once Christiana opened the front door, the first thing that Di'Ban saw was Maximillia; she had crept down the stairs, standing guard next to Christiana; she did not take her eyes off Di'Ban!

Christiana asked why are you over here banging on these doors? Di'Ban looked like a mad dog out of control in the face, and he even had the gall to say that you don't need to be staying over here at some other man's house! Christiana told him, "There is no reason for you to be looking for me or to be concerned with where I'm at. Now, get off Sicarri's property before I call the police, letting them know that you are trespassing!" Di'Ban knew better than to not start any craziness or refuse to leave because Maximillia had begun growling, showing him her teeth!

The following weekend, Christiana went to get her sons and to move them with her! Christiana sadly realized that she was not about to put herself or the boys through what she had already gone through living with her parents. Besides that, her sons were already familiar with Sicarri because they all had spent much quality time together, and he was a friend.

One of Sicarri's neighbors had to have called his ex-girlfriend, whom he missed but could not get to answer any of his calls until she heard that another woman was at his premises.

Sicarri came to Christiana, telling her he needed to talk with her. She and the boys had only been there for about a week before the conversation took place. Sicarri said, "My girl called me yesterday saying she wants to move back with me, so you and the boys will have to leave." Christiana said, "You got to be kidding me, Sicarri. You know exactly what I am going through right now, but OK, I will figure it out, and I appreciate your gesture!"

Christiana called Di'Ban to tell him to come and pick up the boys and their belongings. He assumed that she was coming back also! Once he arrived with a truck and began loading the children's things, he quickly asked, "Where's your things?"

That was when Christiana said, "I never insinuated that I was coming. You are only here to pick up the boys and their belongings, that's all!" Di'Ban could not believe she was not coming because he knew she would never walk away from her children, but this was the one time Christiana was done sacrificing herself.

Christiana called one of her male cousins whom she, at the time, would hang out with, not realizing that he had a sick desire to be with her intimately. She was about to find out how much of an opportunist he was.

Christiana explained to her cousin what was going on, and he already knew what Di'Ban's behavior towards his cousin was like. Christiana was told that she could come there, but the first night of being there with him, he made his way into the bedroom and forced her to lay beside him and would not let her leave as she pleadingly asked him to leave her alone.

Christiana was forced to lie down beside him for a few hours while he held her, telling her that no one would know. Christiana said, "I would know! Now please let me leave," finally, after a few hours, he let her leave!

Christiana loaded her belongings into her car and drove around, distraught from all she had been going through. She finally called Tony later in the morning, telling him what she was going through, but did not expect him to help her.

Tony said bring your things here, and you can park the car in the garage, Christiana went to his house, where they sat and talked! Tony was trying to help calm her down by telling her that everything would be alright and that he would help her put together a plan.

Tony then took her to the grocery store to get some food for the house because he lived alone and did not cook. Tony appeared to come off to her as a trusted friend, which relieved some of her stress, but Christiana was missing her children and could not eat; meanwhile, she was putting money together to get them a place for stability.

Christiana's parents discovered she was no longer living at Di'Ban's and had left the children with their father. Natalie started a rumor saying that Christiana was on drugs and that she had left the kids on poor Di'Ban while he was doing everything that he could for them. Christiana was not shocked because she knew they were all birds of the same feather.

One of Christiana's uncles called her, thinking the rumor was true. As soon as he began to repeat what Natalie was telling the entire family, Christiana said, "I can assure you that I am not on any drugs, and never have I ever been on any drugs!" Christiana also told him, "Besides that, I did not walk away from my children; I left them with their father, who was putting his hands on me until I am able to get us a place to live."

Each day, when Christiana got off work, she would go by Di'Ban's house to check on the kids and prepare them something to eat. Christiana made sure to leave before he would arrive back from work because she had no interest in even looking at him!

Di'Ban happened to arrive back at the house one day before Christiana had the opportunity to leave, and when he walked in, she was sitting on the couch holding Ishmael in her lap. Di'Ban tried to walk up and give Christiana a kiss on her cheek, but she swiftly moved her head away because she did not want him kissing on her nor touching her even with his 10ft pole!

Di'Ban had plans to try and figure out where she was staying, so he went out to her car while she was in the house talking with her sons, then set the meter reader on her car! Di'Ban was trying to find out the range in which he needed to ride to look to find her. Cassius was the one who told his mother what his father was doing because he told him!

Christiana talked with the church minister about needing help to get some money to get her and the kids an apartment! Tony told her that he would also help her come up with some money to help her move along with the minister's help.

Christiana had received enough money to get a place for her and the kids to move into. Tony also rented a truck to help her load all their belongings, and he told her he was not going over to Di'Ban's to get anything. He asked her to see if she could get help from her friends to get all they needed from the house and said that he would have help once she made it to the apartment to unload the truck.

Tony had taken Christiana over to her parents to retrieve some items that she still had remaining in their garage. Once they loaded those items onto the truck, Natalie sarcastically asked Tony, "You're not going over to Di'Bans to move anything. Are you? Because that's not what I've heard!"

Tony turned to Natalie and said, "Well, sounds like you need to consider your source, seeing that he is the same person who threw your daughter and your grandchildren out on the streets!" Tony and Christiana walked off, leaving Natalie standing there looking foolish!

Christiana called and told Di'Ban that she was on her way with a truck to get their things! Once she and a few of her girlfriends had arrived at his house to load up their belongings, he would not let them in to get any of their things!

India decided to talk with Di'Ban to convince him to let Christiana get their belongings. India asked him to sit in his car so they could speak, and that's when he told India that he didn't really want Christiana to leave!

India said, "Now, Di'Ban, you do realize that you campaigned long and hard for this, right? I really don't understand why you would think that Christiana would want to stay here with you. However, you must come to grips with the fallout of what you've created! Now, let Christiana leave in peace!"

Natalie saw India walking down the street the following week towards her house. Natalie goes outside to talk about Christiana to India.

She began saying to India that Christiana ought to be ashamed of herself for how she had treated Di'Ban, but India said, "No, it's actually the other way around, and I don't understand how you would even say something like that about your own daughter!"

Natalie began to play a naive victim role by saying, "Well, I'm only going by what he's told me." And that was when India swiftly said, "Well, there's your problem; what I don't understand is how you could possibly believe him over your daughter! When I know that you have seen his behavior towards her and your grandchildren."

Di'Ban had become so obsessed with trying to get him some get back towards Christiana because he felt that she had bested him! Di'Ban moved "Alley" in with him, thinking that Christiana would be jealous, but to his surprise, he found out that she did not give one good care about who or what he was seeing!

Christiana had gone to pick up Niles one afternoon from his father's house after one of his sports practices, and as Niles made his way to the car, he jumped in with his mother, looking shocked!

Niles slammed the car door, still looking at his mother with disbelief; he said, "Mommy, that woman is so ugly!" Christiana thought to herself that her son was just being loyal! Christiana kept receiving calls from people who knew them both, and apparently, once they had run into Di'Ban and Alley, everyone seemed to have the same shocked responses about how unattractive they thought Alley was, but again, Christiana thought that their loyalties were for her! Christiana would soon get to see what the big rave was all about.

Di'Ban immediately pitted Alley against Cassius because he did not like the fact that he could not get Cassius to turn against his mother! Di'Ban wanted to use his sons as pawns in his animosity towards their mother!

Alley had tested the waters when she tried not to allow Cassius's entry into his father's house. When Christiana had gone to pick up Niles and Ishmael, once, Cassius tried to go inside his father's house to help his brothers get some of their things. Alley told him he was not welcome to come in and then attempted to slam the door on him, but Cassius caught the door and pushed his way inside to get his brother's belongings.

Di'Ban decided to treat Cassius indifferently than Niles. He believed they both were old enough to choose an allegiance and with Di'Ban's delusion of grandeur, he felt they should choose a side. Di'Ban instructed his sons to mess up their mother's relationship with another man, also telling them that their mother's opinion does not matter when it comes to her children!

Cassius comes out to the car carrying his brothers' belongings. Niles is still inside getting the remainder of his things when Cassius tells his mother that she tried to slam the door on him. Christiana goes up to the house to address her behind, but Alley slams and locks the door. Niles said she ran off, then grabbed the phone to call Di'Ban!

The next day, Di'Ban and Alley ran downtown, filing protective orders against Christiana with complete lies about her. When they went to court, Christiana explained exactly what happened and that his father was not even home when what they claimed happened! Christiana stated they could keep the protective order they had filed against her because she meant what she said to Alley if she messed with her kids.

Di'Ban did not ever take an interest in his youngest son, Ishmael, because he was too young to accept their father's delusion of what he thought of his self-greatness! Ishmael would eventually draw to his own conclusion regarding his father by Di'ban's own hand!

Chapter 55: Sweet Treats

Christiana had started and was getting her business up and going with some online dessert orders. She was fortunate enough that she had an opportunity to present some of her products to officials from the governor's office for a ribbon-cutting ceremony.

After the event, Christiana received a call from one of the governor's official wives, who worked for a major pharmaceutical company, and she was seeking out unique gifts to show employee appreciation for the company.

Christiana also received another phone call from a large hotel in the city of Chicago that wanted her services to be added to their dessert menu! The same day after she received that call from the hotel, she received a call from one of her son's schools, stating that he had fallen ill and unbeknownst to anyone, it would be life-changing for Cassius and his family!

Once Christiana arrived at the school, they retrieved her son for his mother, which is when his mother could see something was seriously wrong! Cassius's eyes were flaming red, and he looked as if he did not recognize his mother. She immediately requested that an ambulance be called! Once the ambulance arrived, they did not understand that her son's situation was an emergency!

Christiana's intuition told her otherwise, that this was more serious than what the rescue squad had not taken notice of. Meanwhile, they stated that he probably had not eaten enough that morning! The rescue squad also stated that low glucose can cause you to feel tired and confused! Christiana immediately asked for rescue to take Cassius to the Lurie Children's Hospital in Chicago. However, they stated it would cost her, and she could take Cassius if she liked.

Christiana looked at the rescue squad and then said, "Listen, my son is taller and heavier than me, so if he were to pass out or slump over, there would be nothing that I could do with him!" Christiana went on to say, "Now just take my son to the hospital, as I had originally asked, please, and thank you."

Christiana arrived at the hospital where her son Cassius had been taken; upon his exam with high blood pressure, then a CAT scan report, it showed that Cassius had suffered a stroke.

Christiana was devastated with learning what her son had suffered.

The neurologist placed him on some blood pressure and blood thinner medications and then asked his mother to set up an appointment with his primary doctor upon release. Cassius had regained some of his motor skills by the third day before he was released to go home.

Christiana did as the doctor advised her, and once his appointment had arrived, mother and son were sitting in the doctor's office when the doctor pulled his medical file. The doctor saw where Cassius had recently had a stroke. She then stated, "At his last appointment, it was advised that your son's blood pressure was elevated, so were you all monitoring it?"

Christiana said I don't understand how I missed that, but that is when the doctor said it was his father who brought him in for the appointment. Did he not inform you?

Christiana told the doctor that this was the first announcement of hearing that her son's blood pressure had been elevated!

Christiana could not believe how irresponsible Di'Ban was with knowing that kind of important information. However, she will learn the reason for him being irresponsible!

The doctor advised Christiana to remove all processed foods from the home and feed Cassius a healthy diet. She went home and then cleaned out the cabinets and refrigerator of all processed foods.

Christiana would not get paid again for two more weeks, so she took Cassius with her to see his father and to explain what Cassius had gone through and that she had tried to reach him also! Christiana told him that he did not return any of her calls and that she also needed to get some grocery money to be able to feed Cassius and the other boys a healthier diet! She told him that she did not have any money until she was to get paid again.

Di'Ban stood right before his son, sounding as hateful as he could! Without any concern or hesitation, he said let that nigga take care of them! Christiana told her son let's go! Christiana was not interested in arguing with Di'Ban at a time like that or in upsetting her son.

Christiana then called Tony to ask if he could help her get some food, explaining what had happened when she asked Di'Ban! Tony told her to come to get his atm card to get what they needed! Tony, with a disbelieving tone, said I know he is one shitty ass excuse of a man.

Chapter 56: Sheep's Clothing

Tony and Christiana had been seeing each other for about a year when he asked her to marry him. Christiana believed him to be a trusted friend. Tony was the type of man who opened and closed many of doors for her. He was always ready to encourage or aid her if he heard her speak on something that would enhance her life and her business.

Tony even volunteered to create and set up her website without any questions or concerns about the cost. Christiana was very appreciative of all that he did for her. In return, she automatically gave him 50 percent of her business for believing in and supporting her.

Tony asked Christiana and the kids to move in with him, believing it would save money for her business. Now that they were engaged, it would also allow them to save for their wedding. Tony wanted a big wedding, which Christiana was not interested in, but she knew he had never been married before, and therefore, she didn't want to create any discrepancies about it. Christiana also knew he had not once haggled nor hesitated to help her; therefore, she was pleased to serve him with that same courtesy, but that would soon come to ruins.

Christiana invested her gift of planning and creating a memorable marriage ceremony for them both. During this time, Niles had an opportunity to join a traveling team heading to the Netherlands. However, the trip came at a costly expense. Christiana asked Niles to talk to his father and ask him to contribute half of the trip expenses. Nile's mother replied that she would cover the other half, given her financial constraints.

Di'Ban did not even make the slightest attempt to provide any money to contribute to his son's once-in-a-lifetime trip. In response, Christiana immediately took it upon herself to seek sponsorship for Niles. She made copies of Niles' invitation detailing the expenses and visited places where men typically gathered, such as barber shops, car washes, and gambling spots, to ask if these men would help sponsor her son!

Christiana resorted to making phone calls to collect even more donations for Niles' trip. She also hosted social club events to collect donations but had not met his trip goal. After Tony was involved in a car accident, he gave Christiana his insurance settlement money for Niles's trip expenses instead of using the insurance check for his vehicle repairs. Tony assured Christiana they would continue pooling resources for her expenses because she was unwilling to let her son leave the country without her.

Christiana advised Niles to call his father again because when she calls him, it is always some foolishness, but his father should be able to at least get the athletic gear that he needed for the trip, but his father refused to help. Niles remembered a savings account his father had opened for him years ago, and he wanted to withdraw his money from the account to purchase his needed athletic gear.

Niles asked his brother to drive him to the bank, where he could remove his money, and once Di'Ban found out about his son getting his money out of his own account, Niles' father then ran to a Lawyer and had Christiana served papers for a court day in a Cook County Circuit court.

The legal document requested that Christiana bring both Cassius and Niles with her to a Cook County Circuit court. Di'Ban and his lawyer wanted the Cook County Circuit Court Judge to crucify Christiana and hoped to get her arrested over $ 300 dollars.

Niles knew he had every legal right to go and retrieve his money because Di'Ban had not set up any restrictions to where he had to be with him to be able to remove the funds. Di'Ban and his lawyer tried to convince the Judge that Cassius had gone into the bank pretending to be Niles's adult uncle.

Both boys explained how they had gone to the bank. Niles was required to present his photo ID, and Cassius never pretended to be his brother's adult uncle. The Judge looked at the boys and then realized that their father and lawyer tried to claim this absurd accusation comment to the Judge, who thought they would work to disrupt Christiana and his son's life.

The Judge saw that there was no way anyone would believe that Cassius could pass as Niles's adult uncle because they looked very close in age. The Judge then stated that Di'Ban and his lawyer had wasted the court's time with their ridiculous case.

The Judge told them they needed to figure out what they were attempting to create, but it was better to sit down with his family to work things out. The Judge asked Christiana if she opposed sitting with them and discussing the issues with Di'Ban and his lawyer. Christiana stated she was not opposed to talking with them because she knew her sons had not done anything deceptive or criminally wrong.

Once they all left the court, everyone went to Di'Bans' lawyer's office. They sat in the conference room, but they would not allow Christiana in while they spoke with the children. It became apparent that their only intent was to interrogate the boys more about going to the bank. The boys' consistent statements quickly ended the interrogation motives, making Di'Ban and his lawyer look foolish once again. Niles asserted that he didn't need to steal anything, pointing out that their mother takes care of them. He also told them, "I am sitting here with the latest cell phone in my hand and a nice watch on my wrist, so what's your point?"

Chapter 57: It Never Stops

Christiana was diligently managing her online dessert company. She was busy organizing her upcoming wedding when another challenge showed up as her son Cassius fell ill again. Unfortunately, Christiana hadn't gathered enough money for her trip expenses. Faced with a difficult decision, she had to choose between focusing on Niles' trip or staying by Cassius' side during his health uncertainties.

Christiana made the tough choice and informed Niles that she couldn't let him leave the country without her, citing her unfamiliarity with the people he'd be with and the need to be close to Cassius. Niles was devastated that he would not be able to go. She understood his disappointment as he began pitching and threw a huge tantrum about it! Christiana felt terrible about it, and that's when Niles called Di'Ban to convey the news. Unexpectedly, Di'Ban miraculously found the money to cover his trip expenses, but nothing for his son's trip expenses from earlier as requested, and Christiana was troubled as Di'Ban hadn't attempted to financially help.

After having collected ten thousand for Niles' expenses and needing another ten thousand for herself, Christiana responded with disappointment after finding out that suddenly his father stated that he would go with his son, not offering to reimburse half of his son's expenses to his mother. She also explained and said, "Niles, I have done everything I knew to do to get the money together for you, and the only thing that I did not do, was to sale my ass to make sure I raised enough money for you to be able even to leave the country."

Christiana also explained to him that his father never contributed to his trip. He even dragged them into a Cook County Circuit courtroom over some little funky ass $300 dollar issue while he wanted to go on the trip off of her and Tony's hard work.

Feeling unappreciated for her efforts, Christiana clarified to Niles that his father had not offered a single penny for his trip expenses. Niles' mother urged him to understand, assuring him that another opportunity would arise, given his talents in sports.

In an attempt to secure an increase in child support for her children, Christiana faced unexpected challenges in court. The Judge ordered her to pay Di'Ban's attorney fees without increasing child support. This bewildered Christiana, and she wondered how in the hell did that happen when I am the one with the kids. However, she soon sensed a hidden agenda. Christiana understood that he was seeking a revenge tactic for her not returning to him, attempting to make her life a living hell.

Di'Ban and his lawyer kept dragging her to court. In one court session, Di'Ban audaciously requested the Judge to order Christiana to surrender an heirloom bracelet that was formerly owned by her grandmother, Madeleine. The Judge knew that she was not about to hand her bracelet over. Once the Judge refused to have her hand over the bracelet, they asked for some sort of judgment to settle the fees. When the Judge refused their first request, they sought an order to attach a lean on her home to settle his attorney fees, jeopardizing the only comfort of home for Christiana and the children.

Sheriffs arrived at Christiana's house days later with an order to seize her car. In her absence, Cassius called his grandmother Natalie" because his mother was not at home when they arrived. Natalie came over to Christiana's house. She walked inside and asked the kids where her daughter's car keys were. The boys looked for the keys and gave them to their grandmother.

Natalie casually walked back outside and handed the keys to the sheriffs. Christiana was shocked that her mother would cooperate so easily, knowing the car was her daughter's only means of transportation.

Months later, Christiana discovered that Di'Ban and his lawyer had been paying off the Judge, who was eventually removed from the bench from a reliable source. When Christiana visited Di'Ban's lawyer's office to settle the fees, a paralegal of the Lawyer apologized to Christiana for his behavior. That's when she asked the Lawyer why he was so deeply vested in the case. It was revealed to the paralegal that his actions were personal. Christiana expressed confusion, as she had no personal connection or children with him. The paralegal insisted that it was a personal matter, adding to the perplexity of the situation.

Chapter 58: Bells Ringing

The day before the wedding, Nadia was not her matron of honor this time, and the hostess who helped Christiana get dressed for her first wedding was her Matron of Honor for this wedding. When her dress was taken to the alteration shop, Nadia called, complaining that her train would not be showing once altered. That was when Christiana said, “It’s not about you having a train just like everyone else’s.”

Christiana took charge of ordering and purchasing all the ladies’ shoes for the bridal party to wear at the ceremony. Upon returning home from work one day, she discovered a shoe box on her porch containing shoes that resembled stripper platforms. She already suspected who they were from. Soon afterwards, Nadia called, and while on speakerphone, she suggested the alternative shoes that she thought would look better. Tony intervened, saying, “You can either wear the shoes that Christiana purchased or take a seat with everyone else on the benches.” Nadia remained silent, but she had another plan for the shoes!

On the day of the ceremony, Nadia and her daughter Lynsey were at home going through an emotional roller coaster. By the time everyone was present to get dressed for the ceremony, Ishmael, a young lad, said he heard crying coming from a closet! He then opened the door to see what was happening in the closet. Upon opening the door, he saw Nadia and Lynsey inside, crying and hugging.

Right away, as soon as they saw him, they shouted at him, saying, "Close the door and get out, Ishmael!" That's when he said that he ran to Manius for safety.

Christiana had requested her parents to arrive early at the photographer's request so he could take pictures of the families before the ceremony began. Still, they did not bother to show up for the picture-taking. Christiana again did not see her parents until she was about to walk down the aisle with Niles, who was giving his mother away. Christiana and Tony decided not to take a honeymoon because Christiana was unwilling to leave Cassius out of concern for any possible health crisis. She was nervous that something might arise that she might not be present for since there was no diagnosis for his condition at that time.

Chapter 59: The Quiet Before a Storm

A month after the marriage, Tony's personality began to change noticeably, like a varmint creeping into a barn house! Tony started to show that it seemed like he had some sort of mental issue. Christiana had grown concerned. She started researching personality disorders, suspecting he met the guidelines for a person who could have bipolar disorder due to his erratic behavior. Tony isolated himself in their bedroom; he enjoyed smoking marijuana as if he were self-medicating. After smoking, Tony would reemerge from the bedroom, acting like his old self, offering apologies, and then being ready to interact with everyone.

Christiana knew he was forever floating in and out of behavior changes, and everyone was growing tired of it. Still, there was going to be a reason that kept her tied to him longer than she was going to be needed to, and again, she'd make the sacrifice. Tony had moments where he would be kind and giving, but he could change on a dime and become a complete ass! Tony was very insecure regarding Christiana; at times, he would take his vehicle away, leaving her to figure out how she would get to where she needed to go, but then be mad when she made it out.

Tony also, at times, would lose all responsibility for providing for the family, affecting everyone in the house. Christiana could not believe his behavior towards her and the kids because he knew firsthand what they had gone through before him. Then there came a time when Christiana had to call upon Di'Ban to buy his son's needed school clothes and shoes while Tony was just as selfish in the same regard when both knew that she was struggling to provide financially because she was spending on healthcare costs mostly. That's when Christiana's nephew, who always came through to check on his aunt and young cousins, found out that the boys were in need, and Manius told her not even to worry about it, that he would get the boys what was needed.

Manius and his aunt always looked after each other when it came to being one another's keeper; this was something that the family also did not like. The family decided to scapegoat Manius by untying their love for him because he was close to his aunt. They both realized the real reasons for their behaviors, which caused them to draw even closer. Whenever they would come against Manius, they always found reasons to draw Christiana's name into their arguments with him. The arguments were always due to their narcissistic mental issues of needing to control Manius and anyone else who was close to her. The truth was that they wanted to dictate who and how they wanted him to respect and prop up certain ones, but Manius, being a man with his own decision-making and thought process, always irritated their souls. Manius managed to ask during the arguments, "What does Christiana have to do with what we are talking about?"; he was once told during a disagreement that you two motherfuckers are just alike! During that same argument, Manius was also told by his brother, YaFam, "Fuck both you and Christiana!" Christiana told Manius, "Good to know because I know I have not done shit to his ass for him to feel that way, but he can just stay away from me, period."

Chapter 60: The listener

Christiana had started experiencing insomnia for some unknown reason. She had also gained knowledge beyond her scope about health symptoms despite not having taken any medical courses. Nonetheless, this newfound knowledge helped her keep a close watch on Cassius. She noticed that he always seemed tired and needed to take a four-hour nap daily.

Finally, Cassius made it to his senior year in high school. During his prom, his mother and Tony ensured that he had everything he needed for the big night, including renting him the car of his choice. With graduation day approaching, Cassius was getting ready for his big day, which was the following day.

On the other hand, Nadia's daughter Lynsey, who was also graduating from Walter Payton High School, called Cassius to inquire about the number of tickets he had. Cassius informed her that he had the same amount as she did.

Lynsey said, "Well, I have more people I want to invite, and I figured that you would not need all your tickets."

Cassius told her that he needed all his tickets for his guest. Lynsey became angry and then began taking a head count of who he should be giving his tickets to. Cassius said, "I am sorry, cousin, but I need my tickets just like you do." He told her, "I love you," but before he could finish whatever else he was about to say to her, Lynsey said to him, "Well, I don't love you," and then hung up on him.

Cassius was surprised and then hung up the phone. He asked his mother what was wrong with them. Christiana said, "Listen, son, all I can say is if you sit around shit long enough, you will start to stink!" Cassius said that was a delusional thought, thinking that you don't have people who care about you and that they would not want to be there to support you. However, that was a selfish act on her part.

A few weeks later, Nadia called to talk with Christiana because she needed a listening ear, and she knew her sister was a person with compassion. Nadia told her sister that Damon had left her and was with another woman, hoping that Christiana would talk to Damon for her. Christiana came to her sister's rescue. She called to ask Damon if he could meet with her so that they could talk. Damon met with her as she began to explain that she knew they got married really young, but the grass may seem greener on the other side. Still, if he were to tend to his lawn, he would see that they would be okay. Damon listened but did not really have any response because he was finally living his truth.

Damon continued his relationship with the woman he had left Nadia for. This woman happened to be the same woman that Christiana had told Nadia about several years ago from the Leave the Light On motel. This woman was Damon's true love, and Nadia knew it.

One morning, Nadia called her sister to tell her about a supposed heated phone conversation between her and this woman. Nadia told Christiana, "His whore threatened me, stating she was going to whoop my ass!" Christiana was always more like the big sister, being her sister's keeper. Christiana said, "Wait! She said what?"

Nadia repeated what was supposedly said to her. That was when Christiana said, "Give me her name and phone number, and I will have her info by noon today!" Christiana paused and then continued in a rage, "Nadia, when you get off of work, you need to bring your ass over here and pick me up so we can go pay her ass a visit because you don't let no other woman disrespect you like that. I'll go with you to make sure that no one else tries to jump you. I got your back!"

When noon came, Nadia came calling to see if Christiana had found out where they lived. The one thing that Christiana never thought about was if she was being used as a pawn in Nadia's scheme because all she heard was the supposed threat. Christiana had forgotten that Nadia knew how she was able to get all of Alley's information, and she only wanted to use Christiana for her needs. Nadia arrived to pick up Christiana, and the first thing she noticed was that Nadia did not have any fighting clothes; she still looked like she had office clothes on. Christiana did not mention it because she thought that maybe she just wanted to talk with her. They made their way to where Damon was living with this woman. Christiana told her, "It's that house right there!"

Nadia slowed down, looking up at the house but never stopping. That was when Christiana asked in disbelief, "So, wait, I thought you said that she threatened to whoop your ass, right? I'm confused, so did you just come over here to do a drive-by or what?"

Christiana did not want her sister to look like no scared ass about the supposed threat. That's when she told Nadia to stop the car and yelled, "I'm going up there to ask for her, so come on!"

Once she got out of the car and headed up to the house, she went up to the door and thought that Nadia was right behind her. Christiana got ready to knock on the door. She then turned to look for Nadia. Unbelievably, she had kept her scary ass in her car. Nadia was sitting there in the middle of the street with her foot still on the brakes, with the car brake lights blaring red, which meant she did not even bother to put the car in park. Christiana thought to herself, '*So if some shit had jumped off! What was this bitch going to do, drive off and leave me?*'

The following day, Christiana was working out in her yard along with Tony, when Nadia drove up for a visit. Christiana offered her a lawn seat, and then she sat talking with her. Nadia was discussing her marital problems when she stated that they had an open marriage. However, Christiana did not lay judgment until Nadia said, "As Lynsey said, we are the only ones who were a real family." Christiana made a judgment statement after that comment and said, "Well, aren't you all having a dose of reality because whether a mother or father is or isn't in the home does not change the fact that you are still a family." Christiana continued, "A parent can be in the home and still not be available."

Nadia kept talking; she even told Christiana that the morning of her wedding, she was at home having a nervous breakdown and that it was supposed to have been her.

Christiana said, “Nadia, you’re already married. However, it’s just unfortunate that Damon left you. What are you really saying? Is that no one else is supposed to be married other than you, or what?”

Later that week, Christiana asked her father, Bash, to come over to repair one of her toilets that leaked. Bash also brought Natalie with him. Once Christiana sat down with her mother, they got on the subject of Nadia. Christiana, taking her sister’s side, said, “I can tell she is really having a hard time with the fact that Damon is gone.” Natalie sarcastically said, “Yeah, well, Nadia told me that you are jealous of her.” Christiana saw that her mother was baiting her for some sort of response. Christiana said, “Jealous of her for what? It’s not like she has anything that I can’t achieve or may have already achieved!” Natalie looked like she had just been fed a sour lemon. Christiana then ended the conversation by saying, “Folks, be mad with the wrong person!” which shut down her mother’s negative conversation.

Chapter 61: Darkest Hours

Cassius had contemplated attending an out-of-state school to further his education, but for some reason, he changed his mind. He then attended and enrolled at one of Chicago's Community Colleges. Cassius was excited that he had landed a job that would pay him very well, but a change was about to come, which he was not expecting.

One Sunday, Cassius suddenly developed a hoarse voice resembling a head cold. His mother, Christiana, being a light sleeper, took notice. After about two weeks, the condition seemed to progress into a chest cold. Christiana had grown increasingly concerned after she had observed her son coughing persistently throughout the night. Strangely, no one else in the house had fallen ill.

On a Saturday morning, as Christiana was cleaning the house, she received a visit from Tony's sister. They were engaged in conversation when Cassius's intense coughing from his bedroom interrupted them. Cassius jumped up from his resting bed and rushed to the restroom, where the coughing continued, capturing Christiana's full attention and resulting in her toning her conversation. Tony's sister, concerned, asked, "Christiana, is he alright?" Christiana replied, "I appreciate the confirmation that I am not overreacting!"

Cassius came out of the restroom and made his way into the kitchen. He turned on the water faucet and began drinking water from the faucet, attempting to calm his situation down. Christiana insisted, "You need to see the doctor," to which Cassius informed Christiana, "I already have!"

Christiana asked her son when because she was the one who always made the doctor appointments for her children once they asked her to, and she knew that she had not made anything for Cassius!

Cassius revealed to his mother that he had spoken with a pharmacist who told him to take some cough medications.

Christiana said, "Okay, but a pharmacist isn't a doctor. I want you to go and get dressed because I am getting ready to take you to the emergency room. I believe that you probably have bronchitis or something that is going to require some antibiotics."

Once they made it to the emergency room and as vital signs were taken, unbeknownst to the nursing staff, Christiana had learned to pay attention to blood pressure numbers. Still, she was unaware of what a normal heart rhythm should be!

Christiana questioned the elevated blood pressure reading. The nurse gave her a smart-ass response by saying that his lips and his fingertips would be blue and that the machine must be broken, dismissing Christiana's concerns.

Christiana did not understand how they had a broken machine when it came to taking patients' vitals. Mother and son went and took a seat until they called for Cassius to see a doctor. Once they got into a room, another nurse looked at the vitals. She then decided to verify it personally.

Again, Christiana paid attention to the vitals, and the number had increased substantially. Undeterred, she asked, "Why is that number so high?" Once again, Cassius' mother was told the machine must be broken. Christiana burst out and said that the machine couldn't be broken, and the one out front was broken too. "I want you all to get a working machine for my son, please!"

The nurse kept her cool once she saw the concern while never letting Christiana see that there was something seriously wrong. The nurse immediately had blood drawn and then sent Cassius down for a CT scan. Christiana expected the results to indicate bronchitis or pneumonia, but the reality was about to be much graver.

Cassius was promptly returned after undergoing an MRI scan, and as he was wheeled back in, six doctors walked in behind him. Christiana froze, realizing that the situation was grave. One doctor displayed the MRI of Cassius's chest and abdomen area, and another doctor, without hesitation, asked, "Ma'am, has anyone in your family ever just dropped dead at a young age?"

Christiana looked surprised by the question and replied, "No, why?"

The doctor explained, "Your son is in liver failure based on the blood work, and his liver is so enlarged that it's affecting his other organs. He's retaining a lot of fluid, which is in his lungs, causing the coughing. Your son could drop dead at any moment. It's so severe that we need to hospitalize him before the fluids reach his heart sack."

Christiana had never been as afraid as she was at that moment. She asked, "What are you all getting ready to do?"

Another doctor responded, "We're preparing to hospitalize your son to remove the fluid and for us to understand why he's in liver failure."

Cassius, apparently unaware of the severity, looked at his mother and insisted he wouldn't stay for hospitalization, citing his legal age to sign himself out. The doctors signaled Christiana as if to say you need to handle this situation and stepped outside the room for her to reason with Cassius, realizing he was being foolish.

Christiana tried explaining that he wasn't in a position to make such a foolish decision and that he also heard what the doctor had just said: that he could drop dead at any moment.

Cassius was still adamant about leaving. Christiana was there with him all by herself and was panicking, which caused her to end up arguing with Cassius. The doctors were looking in, seeing that she was trying to talk some sense into him.

Christina took out her phone to call Di'Ban to let him know what was going on and to seek his assistance with talking some sense into Cassius.

Again, Di'Ban did not answer. She left him a voice message explaining how urgent it was, but he did not bother to return her call. Christiana then called Tony, explaining Cassius's refusal and what was going on with Cassius. She told Tony that Cassius had stated that he was not staying at the hospital. That was when Tony said, "Let me speak with him."

Once Cassius got on the phone with Tony, he told him, "I don't know where you think you're going if you leave that hospital, but it will not be here!" Tony also told him that it was too severe for him to be acting ridiculous at a time like that. Tony advised him, "Do as the doctors and your mother are recommending for you because it is a matter of life or death."

Cassius had been hospitalized for four days, and upon his release, Di'Ban arrived. When Christiana saw him, she asked if they could go into the waiting room to talk, hoping to convey the urgency of Cassius' situation. However, as soon as they entered the waiting room, she explained that she had tried to call him because Cassius was in trouble. Di'Ban, harboring animosity towards Christiana, chose not to listen any further and walked out.

The day that Cassius was released, the nursing staff informed his mother that a visiting nurse would be coming to their home daily to administer blood thinner shots to ward off any possible strokes. Christiana took her son home and immediately went into the kitchen to prepare food for him so he would be able to eat a very healthy diet. The doorbell rang, and as she was about to head towards the door, Cassius stated that he would get it!

Christiana went to see who it was. Once she did not hear his return, she saw that it was his girlfriend, and they were about to walk off! Christiana opened the door and asked Cassius where he was going. He looked back at his mother and responded that he didn't want to be there. Christiana stepped outside to implore him to stay. Christiana peered towards the end of her driveway, only to find that her son's girlfriend's mother was behind the wheel. To her dismay, Cassius and his girlfriend were preparing to leave. In a frantic plea, Christiana begged the mother not to take her son anywhere, emphasizing his critical condition due to liver failure.

This girl's mother looked down her nose at Christiana as if she was someone small, from whatever she had been told by either Cassius or her daughter. Christiana was damn near in hysterics, and it took one of her neighbors to calm her down. Reflecting on the situation, Christiana realized her son's frustration stemmed from his inability to control the household, exacerbated by his father's frequent absences. This, in turn, fueled his desire to have his girlfriend spend the night.

After his mother had calmed down, she went over to Di'Ban's house. As she pulled up, Cassius was standing outside with his father and the girl! Christiana again had to argue with her son while Di'Ban stood silent with a smirk on his face, orchestrating the mess.

Once she got Cassius into her car, she explained to him that Tony's health insurance was paying for his healthcare, and everything was registered to their address. On his third day home, he wanted to visit his father's house. It was his friend Amir who convinced Christiana that he would be with Cassius and assured her not to worry, as he would call her if anything seemed wrong.

Against her better judgment, she let Cassius go. Just as Christiana began to drift off to sleep around 2 am in the morning, her phone rang. Amir said, "Mama, Cassius just had a stroke. The medics are here and want to know what medications he's on."

Amir then reported the routine. They were transporting Cassius to Mount Sinai Hospital, and that, as if to relieve his mother of any undue stress, reassured her that he would bring Ishmael with him. Christiana, concerned and frantic, inquired about the father's whereabouts in such an emergency, but Amir's response only worsened her stress. He reported that he didn't have the slightest idea where he was and that he was gone before the decision to move Cassius to a hospital was discussed or considered. As if desperately looking for some solution or relief, Christiana asked where Niles was, but Amir responded helplessly with the same cluelessness, saying that he didn't know that either.

Christiana lived a considerable distance away from the hospital, but for reasons and skills only a mother would know and could manage, she was able to beat the ambulance to get there. She was pacing restlessly outside the emergency ambulance entrance, waiting to see her son once they pulled in. Amir and Ishmael were also already at the hospital.

Thirty minutes later, Di'Ban strolled in with some woman. In all this chaos, he had taken the time to go and retrieve a lady and even bring her there instead of racing to the hospital for his son! Christiana was fuming. She asked him, "Where's Niles? and his response was almost expected, and yet all the more infuriating, he said, "I don't know!"

Christiana failed to make sense of or even comprehend how irresponsible he was, wasting precious and crucial time retrieving a female before going to the emergency with his sons and not even knowing the whereabouts of his 15-year-old child!

The following day, Christiana arrived at the hospital to check on Cassius. His girlfriend and her mother were already there and displayed nasty and uncalled-for arrogance towards Christiana!

Christiana couldn't have cared less. The only person she was concerned about at the hospital was her son! She immediately checked in on Cassius and asked if he wanted anything, maybe from the gift shop. Cassius asked for a few things, and Christiana returned with them in no time. By that time, Di'Ban had arrived, too, and all that were in attendance greeted her with cold welcomes. Christiana understood the foolish arrogance of them all!

She decided that it was better to leave before she stooped to their level. When she got home, she packed up Cassius's belongings, all of it because she didn't want her son anywhere he didn't want to be or where he would feel neglected or carelessly attended to. She asked Niles to put all his brother's belongings inside his vehicle and then take the car to their father's house right after he was released from the hospital.

Soon after being released from the hospital, the health insurance company that Cassius and his brothers were covered under, Tony's Insurance Plan, came up with a rather unusual concern. They reported that the nurses were going to another address other than the one that the insurance policy had listed for.

Christiana thought that they were trying to cancel him off insurance. She informed them that he still resides with her at her address, and he was just visiting at his father's temporarily.

She let her home-based business lay dormant for a while. She was already dealing with the stress of the unknown with her son and with Tony's bipolar personality. But that couldn't have lasted for long because Christiana had to work to ensure a steady income to keep the hopes of moving out. Little did she know that the unknown was about to make an appearance and disrupt their lives tragically.

One afternoon, while at work, she gets a call from Cassius. He sounded terribly sick and informed her that he wasn't feeling very well and needed her to come to take him to the hospital immediately. She asked him whether his father was around for that, to which Cassius informed her mother that he was but wouldn't take him to the hospital!

Christiana assured her ailing son not to worry as she was on her way and would be there in no time. However, she had to call Tony to pick her up from work and take her to transport to take Cassius to the hospital! When she reached their father's house, he was there but didn't even bother or attempt to try and walk Cassius out to the car despite seeing him in frail health and struggling. Once Cassius was out on the porch, Christiana and Tony had to get on both of his sides to be able to get him into the car.

Fortunately for Cassius and his mother, the visit didn't require a hospital stay. Christiana, a bit relieved, took him home with her. He was taken care of and attended to until he recovered. When he was up and healthy, he returned right over to his father's house. That was a mistake on his part, a very grave mistake.

Roughly about a month later, Christiana was home when Cassius came for a visit! He walked, or what could only be called dragged himself, from the driveway to the front door, which was a short walk for anyone in normal health. Once he came into the house, Christiana's heart wrenched as she saw how worked up and exhausted her son appeared. He immediately had to take a seat. Christiana also noticed that her son was diminishing at an accelerated rate along with helplessness, then pain further settled in again, and it wrenched her heart. She did not know what to do. But there was more to it, and she was about to find that out in the most unbelievable way!

Christiana, by practice or habit, had a couple of things that she would always notice when she met people. She'll remember the eyes or the smile, depending on which one hooks her attention first. There was a lady that she had met through one of her friends. It was a one-time visit, but when they met, it was the woman's eyes that had lodged in her mind!

Christiana was never the type of person who would pry into anyone's business. However, one day, on her way back to her home she stopped to get gas when she saw the same woman that she had met through her friend. The woman had a surgical mask on, but Christiana recognized her instantly because of her eyes! She decided to speak to the woman who failed to recognize her at first but remembered instantly when Christiana reminded her about how the two had met through a friend.

After a brief chitchat, Christiana developed some courage to ask her why she was wearing a surgical mask. The woman, against Christiana's expectation, did not find the question offensive or out-of-line, which was quite relieving for Christiana. Instead, she casually said, "I just had a liver transplant," a revelation that widened Christiana's eyes! The woman then went on to narrate a short run-down of what had happened to her, and Christiana thought it would be a good window to share her own plight with the lady. She said, "Listen, I am going through something right now with one of my sons, who too has a failing liver, but to what extent, we are not sure of yet!"

She filled her up with more of what was going on with Cassius. The woman listened intently, and when Christiana was done, she instantly handed her phone number to her. She advised Christiana to call her as soon as she got home, and that was exactly what Christiana did. They talked for a considerable while, exchanging more details and information. During the call, the woman provided Christiana with a list of doctors to reach out to and gain more insights and medical assistance on Cassius's liver issue.

Talking about her son's health took a heavy toll on Christiana. She felt like she was losing her mind out of fear and all the terrifying thoughts about her son not surviving his failing health! She wouldn't pay much attention to whoever was around her, and the thought of getting her son back to good health was the only thing that would keep her occupied, even at work. One day, while Christiana was at work, unbeknownst to her, there was a lady who had been observing her from a distance for quite some time. The lady noticed that Christiana would stay to herself all the time, not even leaving her desk for lunch.

One particular day, when everyone was away for lunch, the woman stayed back, and when she found no one around, she walked up to "Christiana." She carefully chose her words, not to sound scary, and said, "I don't know what is going on with you, but I can tell that your heart is heavy, and I want to pray for you right now!"

Christiana, as if longing for such a gesture of concern and care, instantly broke down. Words came out without much effort as she informed the lady, "I have a sick son, and I don't think he will make it!" She wanted to share more but knew that she didn't have the luxury of breaking down for long and collected herself with all she had in her. Right then, she nudged herself out of her emotions and decided to go home without giving any explanation to the woman, but she knew the concerned lady would understand a mother's emotions. She arrived home tired and without an ounce of energy or will to do anything, not even to eat. She went straight to bed, completely clothed, even with her shoes still on.

But what all looked random and chaotic and tiring for her was just to call her to something more surreal. While she was asleep, the "Most High" spoke to her. He mildly reprimanded her and told her that He couldn't get her son's attention because she kept running up and getting in His way! The "Most High" reassuringly told her that once she got out of the way, He would get her son's attention!

The following evening, Christiana received a phone call from one of her relatives. She informed Christiana that her father Kingston, who was Bash's brother, was ill with cancer and that Bash needed to get there because she did not believe he would hold on much longer.

Without wasting another second, Christiana called her parent's home. Her mother, Natalie, answered the phone, and without much preamble, she asked if her father, Bash, was there and that she needed to talk to him urgently! Bash took the phone, and Christiana filled him in. There was worry, concern, and urgency in her voice as she said, "I just received a call that your brother, Kingston, is ill, and from my understanding, they are not expecting him to survive much longer." After a detailed exchange of information and reassurances, Christiana said,
"I can take off tomorrow, or we can leave tonight if you like. I'm willing to drive you!" Her father agreed, and Christiana rushed to pack her a bag. She had to be ready to travel with her father from Chicago to Wisconsin because it demanded urgency nonetheless. Once she had her stuff ready, she went to ensure that her sons looked after each other. She went out to look for Niles and wanted to instruct him to look out for his younger brother, Ishmael, while she was gone.

Understandably, Christiana was up later than expected by her family. After rummaging through the house and not finding him, she went outside to look for Niles. She found him and a friend sitting in her car. The two young men looked to be in deep conversation, and Niles was visibly sad and melancholic.

The two were engrossed and didn't see Christiana coming. She knocked on the passenger car window, which surprised them both, partially because of the sudden tapping on the window and partially to see that she was up late. A bit alarmed and alert, she asked them what was going on. The friend dropped his head and said, "I don't know!"

Christiana steered the question towards her son and asked, "Niles, what was going on? You both looked as if you were in deep conversation!" Niles dismissed the interrogation and winded up the deep conversation. The two friends decided to part for the evening and did so in a hurry that Christiana couldn't interrogate any of them as both got out of the car in a flash. Once Niles came into the house, he spoke before his mother would ask any further questions and told her that he needed to tell her something!

With words that felt heavy and bitter, Niles weakly said, "Cassius does not think that he is going to survive and that Di'Ban had been leaving him at home by himself throughout the night!" Niles took a pause, which wasn't a longer one, and then went on. He informed his mother about the dread her ailing son was going through. He narrated that his brother told him that he was afraid to go to sleep at night for fear of not waking back up! Collecting himself, he added, "I have something else to tell you… I have been taking your car and sneaking out of the house at night once Tony leaves to go to work, and you are asleep, to go over and be with Cassius at night!"

Niles didn't want to worry his mother and thus was finding it very difficult to say the words. After a very brief pause, he went on, "I see that I did not have a choice this evening," and further added, "Cassius did not want me to tell you! Cassius also told me that if he does not make it out of his health crisis, what he wanted each of his brothers to have of his things as a reminder of him."

Then he went into the infuriating bit of information. Niles told his mother their father had gotten Cassius to sign an insurance policy and had pressured Cassius to sign it. All Christiana could say, seething inside and out, was, "That is amazing!! He made sure to get a death insurance policy but refuses to add his children to his health insurance plan!!"

Christiana called to talk with Tony about the latest revelations about Cassius but Tony wasn't much help. He only managed to say what wasn't helpful to Christiana one bit as he said, "Well, it was foolish on his part for even going over there. I know that they are your kids, but trust me when I say that blood will come out!" He further added, "You will eventually see for yourself because I have already told him that he needs to stop lying to you, and I know that he didn't like it, but it's true, and he is listening to the wrong person in the first place."

That same evening, Bash called to tell his daughter that he was on his way. She asked if there was anyone else who could drive him because she was now dealing with something concerning Cassius. He told her that "Nadia and Andre" were not prepared to take him. Christiana, exhausted and without much energy to protest or even hold discussions, told her father to come and that she would take him!

Once Bash got to her home, they hit the road instantly without any unnecessary delays. Making small talk, her father said, "Your mother is upset with you!" Christiana thought that was nothing new but then asked, "What is she upset with me about?" He informed her and said, "You did not tell her first about the phone call conversation that you received..." Christiana interrupted him, mildly irritated but mostly because she knew where the conversation would go, and said, "It was hard enough delivering the news to you, and besides that, I did not want to keep repeating myself!

I knew that you would tell her after we got off the phone what was going on."

Bash still continued and said, "… and your mother was wondering how come they did not call Nadia?" Christiana, vexed and out of any patience to hold herself back, retorted, "What does it matter who they called? Listen, all I know is that they called me for whatever reason, but who called who should not even be an issue. Now, if my mother wants to know why they did not call her other two children then she needs to call and ask them for herself because that's crazy to even squabble about!"

Christiana dropped her father off at his brother's to spend some time with him, and she arrived back home the following day. She got up early the next day and called the health insurance company to see if she needed to wait to switch service providers. To her surprise, she found that she did NOT need to wait or get a prior authorization.

Christiana called Mount Sinai Hospital and was able to make an appointment for Cassius once they understood how severe everything was. They confirmed to see him within a few days. The Gastroenterologist and Hepatologist scheduled a battery of tests on Cassius within days, and the physician then made a return appointment for Cassius.

Upon the next appointment, Cassius and his mother were informed that he had a genetic disorder that was going to require a transplant! Christiana watched her son Cassius look like he was having an out-of-body experience upon hearing the word "transplant!" The physician asked Christiana if she had any questions, and the only thing that came to her mind was to say, "Please, save my son!" But that was enough and even understandable for the physician who reassured them and said he understood because he also has children and that he would do everything within his power to save her son.

The physician then put the doctor's cap back on and got into the details of the operations. He told them both that the process was going to be rigorous and that Cassius would have to endure, but it was necessary in order to be placed on the transplant list!

Christiana had received court documents declaring that their father was attempting to emancipate Cassius and all legal responsibility concerning his son! Christiana had to prove that Cassius was what is known as a DISABLE ADULT CHILD to have the courts stop his father from emancipation, and it was also going to aid Cassius to remain on Tony's health insurance due to his age and the fact that he was no longer a student.

Christiana, without being given any choice or considered in any discussion, was declared as a caregiver for her son Cassius! She had to stop working in order to ensure that she was able to take her son to each appointment most days, and it would be 3-4 appointments a day each week, Monday through Friday. She knew that it was not an option nor a luxury for not being there, unlike their father, who chose not to be there during his son's dark journey! She did not once receive a call from her immediate family or Di'Ban asking if there was anything that they could do for Cassius or for her. Christiana even talked with their father one day about how tough things were taking care of him and the other boys. She asked him if he could at least send her an additional twenty dollars a week. Needless to say, she never received any help from him.

One day, while speaking to a relative, she was expressing how unbelievable of an ordeal it was for her as a mother to watch her child go through such a fight for his life. Her relative had the audacity to give the most bizarre response as they suggested to "minimize the seriousness of what Cassius was enduring." Christiana was coming up to the realization that she thought she had an inner "circle," but that circle would soon become a dot!

Ironically, it was not the immediate family who stepped up to the plate to help them, but some cousins who were not even raised together. They showed up to help her out of nowhere and came out daily for 2 ½ years, consistently without stopping once! The "Most High" knew exactly who he needed to send. It was magical and miraculous for Christiana. Those family members had no clue of the level of appreciation and gratitude Christiana had for them!

During all this time, the only good grace was Ishmael. During the entire waiting period, he was always on standby, being asked to run to get anything that Cassius would ask for, even if it was to clean his car for him so that he could take a ride. Ishmael, at a young age, understood what it meant to be his brother's keeper without question or quorum! He always made sure that if no one else asked how his mother was doing, he would because he was the one who watched her go through it all!

Tony realized that Christiana had a large amount of money, but due to his sporadic bipolar behavior and sporadic selfishness, she did not mention it to him. She had begun looking for a new residence to move her and the kids into because it had been too stressful with everything that was going on. Not only was she dealing with a sick son, but she was also dealing with her middle son trying to run wild and her constantly having to talk him down! Then there was the angry, jealous ex who was trying to make her life hell and a husband who was changing on her daily. Then there was the youngest son who was scrambling to find his way because no one took the time for him other than his mother, that too when she was able to fit a few minutes in her day just for him. And not once did anyone ever think of all of this other than that youngest son.

One day, Tony wanted to argue with Christiana, but she would not entertain the nonsense, so he decided to go for what would rile her up easily. He started with Cassius and began with his usual aggressive behavior towards her son. However, much to his surprise, it escalated into something that he did not realize would ever happen.

Just when he started the physical altercation, which would usually lead to physical violence, the police were called, and they came and arrested Tony. Once the bipolar husband was out of the picture, things smoothed out a little. Cassius was hospitalized, and the process to take the stress off his liver to increase his chances of being transplanted much quicker began shortly afterwards.

Chapter 62: Where Your Ass Was At

Miraculously, the same people who had not bothered to call or check on Cassius before the altercation had taken place were the first ones rushing to the hospital afterwards. They were only there to add layers of bullshit on top of an already messed up situation.

Christiana had already expected that they would go jumping on a drag wagon. She was well aware of those folks who had not done shit and were not going to do shit but were quick to run their mouths like a pack of useless hyenas.

Di'Ban also told Cassius upon his visit to that drag party that he would come for his son's transplant as long as Christiana was not present. In response, Christiana told her son, well, Di'Ban's self-loathing comment was ignorant, and she assured him that he wouldn't be present.

Despite standing by Cassius as she always did, Christiana noticed that he was back to acting foolishly and unappreciatively towards her. His mother understood that if one sits around shit and serpents long enough, then they are going to either start to stink or hiss accordantly with spewing poison. Christiana consistently emphasized to her sons that the first line of attack from such individuals would be on their reputation, especially for those who were foolish enough to lend them an ear.

Christiana asked Cassius what was his problem with her. He bluntly told his mother that she was selfish.

In response, Christiana said, "Okay, well, I see where that's coming from." although she remained composed and calmly invited him to discuss the accusation, she said, "Let's talk about me being selfish. Now, you can still call me selfish if you like, but let me tell you what this selfish mother has done for you." Christiana continued, "Not once have I ever left you. I have been by your side throughout everything in your life. I sacrificed myself for you and your brothers, often going without to ensure your well-being. I shared everything I had with you and your brothers, even if it meant wearing raggedy underclothing to make sure that you and your brothers had underclothes. I have sat alone at hospital bedsides by myself for you and your brothers. I have been at your and your brothers beckoning calls no matter the day or hour. I chose to be your mother, doing everything necessary to take care of you and your brothers other than to sell my ass. I have given when others would not even give you or your brothers the sweat off their brows. I have given without even placing myself in that equation! I have even risked my life for my sons. I stayed in two of the nastiest relationships for you and your brothers, and I have gone to war for you and your brothers! I know what I have done, but I also know what I have not done; now, if my character has been judged as selfish, then so be it!"

Cassius instantly started to talk about Tony, and that's when Christiana said, "Listen, son, I care, but I don't care about what happened. What I care about is that you live because he had not gone and dropped that health insurance on you, and right now, you need health coverage more than ever before, so if I stay with someone for that reason alone, then so be it!"

"Cassius, without that health insurance, they would ship you off to the county hospital, and trust me, I know you would not survive." Christiana continued, "Besides that, without that health insurance, you would never be able to recover financially! Cassius so rests in the fact that he got what he had coming, but you also need to remember not to listen to those who have not even given one red cent towards your health care nor cared to partake in any part of your journey! Cassius, I need you to understand that the cheerleading squad can say whatever they want to say about me, but what truly mattered to me was that they shouldn't have ever parted their rotten lips!"

Christiana and Ishmael were faithful and loyal to no fault, as usual, for going to the hospital several times throughout the day. Christiana would go early in the morning to collect her son's undergarments and then would return home, making sure to do his laundry. She occasionally prepared healthy meals for her son to take back to the hospital and for the nursing staff to show appreciation towards them for taking care of her son.

Once Ishmael arrived home from school, they would prepare to go back to the hospital. Ishmael always prioritized visiting his brother while walking the halls with him to provide support and show Cassius that he was not alone in his ordeal. Daily, Ishmael would ask his mother to stop to pick up something for Cassius, be it from the gift shop or from the mall, even if it was gummies or popcorn, as long as he brought him something and made him happy.

Ishmael consistently saved any money he received, and without hesitation, he would ask his mother to take him to a hobby shop to purchase his brother's model home kits so that he would have something to occupy himself with while awaiting a liver.

Christiana knew how selfless Ishmael was when it came to his brothers and noted how he never thought of himself. The hospital staff grew fond of Ishmael recognizing how loyal and supportive he was as a young brother, and these nurses all treated him like a little brother, appreciating his positive presence.

On one occasion, Christiana went to one of her neighbors to check if Ishmael could continue to swim and stay with her children. This neighbor said that Ishmael was fine with staying with them and that she would look after him until she arrived back home.

Christiana then went to let Ishmael know that she was getting ready to go to the hospital and that he could stay to swim with his friends, but Ishmael sprung up out of the pool and said, "I'm going too!" Ishmael insisted that he could swim anytime, but he wanted to be by his mother's side at the hospital. Ishmael's mother was shocked to see that even at that moment, he still had his brother first on his mind!

Christiana and Ishmael always had the same conversation on their usual rides, going out to the hospital and observing motorcycle riders without helmets. They both questioned and wondered if any of them would be potential donors because the transplant team had told them those people were usually the donors since these riders seemed to undervalue their own lives.

Christiana had a friend that she had grown up with "Micah" who worked for the hospital. He knew that Cassius was her son. Micah took the pleasure of turning Cassius's room into what looked more like an apartment so that Cassius would not grow depressed by the hospital environment.

One evening, when Ishmael and his mother were about to leave from visiting Cassius, they were approached by a family member of another patient. They inquired about how Cassius was holding up, and before the conversation came to an end, she asked Christiana if she could ask her a question.

Christiana answered her and said, "Of course!" That was when the family member asked her, "Who takes care of you?"

Without haste or hesitation, Christiana pointed towards her youngest son, Ishmael! That's when the family member said, "That's unbelievable for a child to have that much compassion,"

Christiana smiled and said, "He has always been a blessing, and he is one of the most selfless people that I know, and I'm not just saying that because he's my son but because it is genuinely true."

One evening, Ishmael noticed the fatigue etched on his mother's face after she returned home. Seeing her looking for something to eat, he immediately told his mother to go rest and that he would fix her something to eat. Christiana, realizing his determination, relented, and he prepared a grilled cheese sandwich just the way she liked it.

After several more months, Christiana stopped discussing the transplant because she had begun to lose faith that her son would receive the gift of life within time to be able to save him. However, she never stopped praying. Christiana had begun to bargain and pray much like her grandfather "Jacques" had done regarding her grandmother Madeleine; she also stopped preparing herself for the next day upon returning home from the hospital.

Chapter 63: In His Timing

Christiana had gotten up to see Ishmael off to school as usual. She retreated to her resting bed, and as soon as she had drifted back to sleep, her phone rang. It was Cassius!

One thing about Christiana was that she always answered her phone, at least by the second ring, when it was her children! Christiana saw that it was Cassius calling but figured he was just calling to talk to her as he usually does. She answered, "Hey, son!" Cassius asked, "Were you asleep?" "Nope, not at all." She said. Cassius asked his mother, "Guess what?"

Christiana could not get her thought process together fast enough before Cassius said it was something that she had been waiting on. However, she still couldn't think. She wasn't prepared that the news was going to come to her in the manner it had because she had been given a pager for when an organ had been located for him! "I don't know, son. What is it?" She asked.

Cassius said, "Mommy, they have a liver for me, and they just came in to tell me!"

Christiana was now fully conscious and stunned; therefore, she tried to remain calm in the fear of making Cassius nervous! She knew she sounded calm to her son, and she then managed to say, "Okay, son. I am on my way!" Christiana sprung up out of her bed like a rabbit. She immediately ran to where Niles was in his bedroom and told him they had a liver for his brother!

Christiana asked Niles if he was leaving with her, but he told his mother that he would come with one of his friends. Christiana did not ask another question. She went rushing to get dressed and became very anxious as her mind raced while looking for garments that she could not locate, not realizing that she was standing on them. Christiana did not take the time to shower or brush her teeth. She was running from room to room because her memory of where she had left certain items like her purse and car keys was drawing blank!

Christiana drove to the school to retrieve "Ishmael." Once the school retrieved him, as he walked into the office, he saw his mother's eyes were large! Ishmael asked, "They got him a liver?" Christiana told her son, "Yes, they do." And the staff told them that they would keep them all in prayer!

They hurried to the car, and that's when Christiana remembered that she had left the pager. She stopped and put her car in reverse gear. Ishmael told his mother that she was doing sixty in reverse! Once she picked up the pager, she rushed to the hospital. In the parking lot, Ishmael laughed and told his mother that the car smelled like hot rubber in an attempt to try to distract and slow her down because he could see she was moving at warp speed! They made their way to Cassius's room. Christiana saw her son was in deep thought from the way he was staring at the floor! Ishmael walked in confidently and said, "I knew it was coming. I have been praying hard for you, and I felt it was near!"

Christiana sat down, hoping to ease her son's fears. She asked her son to sit with her as she began to tell him that no matter how old he was, he was still her baby! She told her son that it's alright for you to feel whatever you are feeling. She understood she had to become a coach from time to time to keep her son mentally strong. She also knew that he would feed off her energy; therefore, she needed to remain confident in front of Cassius, showing him that everything was going to work out!

The transplant team soon arrived to talk with Cassius and his mother to explain what was about to happen prior to surgery. The first thing that they were getting ready to do was to suppress his immune system before the liver arrived!

The transplant team also told Christiana that they understood the people were getting ready to want to come for support, but she needed to let them know not to come if they were sick and to not bring any children because they carry childhood diseases with spores on their clothing, and even if they were not sick it could kill her son.

The team immediately gave Ishmael fresh scrubs to change into. They then had him scrubbed down only because of him being Cassius's brother and knowing that he was not ill, but that was the only exception that would be made regarding children being present.

Christiana soon began calling the family, letting them know that Cassius was being prepped for the surgery and that he would be going into surgery sometime soon. The transplant team repeated if they wanted to come to the hospital, it was fine as long as they were not sick, but if they were sick, they shouldn't come! Christiana was again told to let everyone know they couldn't bring any children with them. The team knew his mother understood thoroughly, but people would try even after being told!

Christiana needed her pastor to be called because her pastor stated that he wanted to be there to talk with her son and to pray for Cassius before he went into surgery! Christiana gave "Nadia" the phone number and asked her to call that pastor for her. Nadia never made the call to "Christiana's" pastor but chose to call her own pastor! The Pastor who was asked to be called eventually told Christiana that he never received any calls from her sister.

Once Nadia arrived at the hospital, she stated that Lynsey was with her and had brought her children, but they were in a waiting room that was down the hall from Cassius's room. Christiana told her that they could not come in because his immune system had been suppressed, so make sure that they didn't come down here!

The transplant team had gotten ready to transport Cassius down for the surgery. The transplant team struggled to get through the hallways because the halls were lined with people who were there to support Cassius!

Lynsey came out of the waiting room and made her way down to where Cassius was standing. She stood beside him along with her children! Cassius looked at his mother, and she sternly said, "Son, come down here now!" Lynsey handed her baby to "Niles," stating to "Niles," "He looks just like you." Niles looked up and saw his mother looking at him, but he understood right then that he needed to give the baby back to its mother.

Without hesitation, Christiana said, "Niles, hand that baby back right now!" Nadia apparently became offended by her response, then quickly jumped with a sarcastic attitude and said, "What? You are not ready to be a grandmother!" Christiana was one to address all questions or comments whenever they were directed at her! She immediately said, "Well, since you asked, but no, I'm not ready to be a grandmother, and nor should you at the age of 40!"

Natalie and Nadia both felt some type of way about Christiana's response because they decided to try to make Christiana feel like she was less than due to her response. All that they had left in them was to immediately reference her by calling her "Gia"! This was due to their lack of liking and love for Bash's sister because she never had grandchildren!

Once the transplant team gathered Cassius onto the transporting bed, they explained to Christiana that only immediate family would be allowed to go back to a nurses' station just outside of the operating room to spend a few more minutes with Cassius before they wheeled him into the operating room! They also stated there was a waiting room where everyone else would be able to wait, and it was on the same floor. Once they made it back to the operation area, "Christiana's" emotions got the best of her. At the same time, she held on to her son while standing at his bedside as the tears rolled uncontrollably down her face until the transplant team told her that it was time for them to take him into the operating room.

Christiana said to her son, "I will be here waiting for you when you come out, and that the "Most High" was with him." Christiana did not know if that was going to be the last conversation that she was to have with her son. Cassius remained calm and said, "I'll see you soon." And that walk back out to the waiting area seemed very long to her!

Christiana made her way back to the waiting area, and that was when the family began their antics of nonsense, and not once did anyone ask "Christiana, Niles, or Ishmael" when they came to the waiting area how they were feeling. Christiana knew that it was going to be a lengthy surgery. The first thing that she did was check with the nursing staff to get blankets and pillows for everyone! The nursing staff also gave plenty of snacks to everyone because they respected Christiana for being the faithful mother that they had seen her being.

Andre had finally made his way to the hospital, but by that time, Cassius had already gone into surgery! Christiana walked outside to get some fresh air because she felt overwhelmed by emotions. Andre followed her and asked, "Christiana, let me ask you a question. What happens if the organ doesn't work?" Andre's girlfriend looked at him and said, "Andre!" out of complete disbelief.

Christiana's brain felt like it spun. She looked at Andre with a nonblinking stare while thinking he got to be the most stupid ass there is! Christiana quickly asked, "So what do you think will happen?" Andre played an ignorant role by saying, "I didn't know!" Christiana turned and walked away, realizing that she was dealing with someone who did not have the common sense to understand the obvious! Andre left the hospital after his comment and did not return!

Christiana arrived back to the waiting area, talking to those who cared to talk to her, while at other times, she sat quietly to herself, praying.

Periodically, one of the transplant team nurses would come out to let her know where they were at with each surgery stage, and around the 6th hour, the nurse came back out to give her an update that Cassius was maintaining well; however, he was still in surgery.

The liver had arrived at the hospital. The nurse came back out, letting Christiana know that they had the liver and that it was about to go into Cassius! Once more, the nurse came out and asked Christiana to come with her because the surgeon needed to speak with her!

Christiana was taken to a private, quiet room where Bash and Natalie" followed in behind her. Christiana took a seat beside her parents. Meanwhile, no one was saying anything. They just sat in silence, waiting! Christiana was going to look at the surgeon's face because she was going to try to brace herself for a worst-case scenario due to them taking her to a quiet room!

The surgeon walked into the private room, but Christiana could not get a read from his face! The surgeon pulled up a seat directly in front of her as they both looked eye to eye. The surgeon placed his hands on her knees. He could see that she was terrified! The surgeon then said, "First, let me tell you that your son is very amazing, and the liver is in; however, he is being stitched closed in surgery right now!"

The surgeon said I walked away because I wanted to inform you myself of how well he's doing now. I need to prepare you for what you are about to see next! Christiana, your son will remain in a medically induced coma for a couple of days until we can calm things down! The surgeon went on to say, "You are going to see a lot of machines and medications hooked up to Cassius, and he will be on a ventilator. However, there will be nurses with your son constantly." The surgeon said before he exited the room. "Congratulations, because I know that this has been a stressful ordeal for you all, but know that your son is in the heavenly father's hands!"

As soon as the surgeon walked away to leave, Christiana sat there staring at the baseboard, completely mindblown, thinking it was really over. It felt like a blink of an eye to her! Bash seized the moment. He opened his mouth and said, "Christiana, you need to find a better way of talking to Lynsey."

Christiana looked at her father. She then took in a deep breath and said, "You know what? I don't need to find anything because that's my child who's lying back there with his chest lying wide open, fighting for his life! I told everyone not to bring children up here. I also told Nadia. Besides that, I never said anything directly to Lynsey!" Christiana went on to say once more, "I do need to find anything!"

Not once did any of the family take the time to say "HalleluYah" or let's go into prayer and rejoice, but the nonsense was about to rear its ugly head shortly afterwards! Later in the day, right after the transplant, some of the other previous transplant patients who had heard that Cassius had been transplanted began calling the hospital or visiting the hospital to lend support through their experiences to Christiana for Cassius's sake!

There was another mother who had come to the hospital because she already understood the unknown right after the transplant due to her daughter being a transplant patient! This mother was sitting and talking with Christiana, and you could hear that they were in deep conversation about what she needed to expect once Cassius returned home and to tell his mother that it was another journey he would endure.

Natalie was sitting and listening to the conversation between both mothers, but she could not hold her sarcasm any further because there was no attention coming her way! While the visitor was still talking, Natalie blurted out to the woman, "Well, how you doing?"

Christiana looked at her mother as if to say, damn, you just can't keep from trying to start some shit even at a time like this. The other mother looked so confused by Natalie's behavior that she could see that it made Christiana mad!

Once Cassius had regained consciousness and had gotten to an acceptable stable level, he transitioned to the level of going home. After a couple of weeks, his mother was given a list of what not to do as well as what they could do! The first not to do on the list required not allowing people to come over nor letting people run in and out of the house due to him being immune suppressed.

Chapter 64: It's Showtime

Natalie and Bash became upset because they were also not allowed to come over after Cassius arrived home from the hospital, but Christiana told them that it was nothing personal and that no one was allowed to visit for a while. Christiana also explained that they themselves try to stay away from Cassius as much as possible and that they should not be offended.

Cassius had become bored after several weeks of being stuck in the house, and he wanted to get out to take a ride and see a different scenery. He drove himself to visit his grandparents so they could see how well he was doing.

While he was there in their presence, he masked up and made sure to keep a distance! While he sat with his grandparents, Bash asked Cassius, "What has your brother Niles been doing?" and Cassius asked what he meant by his question.

Bash responded with a comment that did not sit well with "Cassius!" So he said, "I see your brother going to jail!"

Cassius asked, "Why would you say that granddad?"

Bash had a cunning tone and then said, "I know people, and I hear things!" Bash made sure to say before he finished his inquiry regarding his grandson, "This is between me, you, and your grandmother, but don't tell your mother!"

Two months later, Christiana wanted to give it a trial run to go back to work for a few hours a day. She arrived back in from work one afternoon and then saw she had a voice message on her phone. Christiana retrieved the voice message, and it was from her mother "Natalie," stating that they had received a phone call from some man stating that Niles had stolen from him and that the caller also said that he was going to kill her son and blow him away!

Christiana understood that just as you can tell when a child is lying to you, you can also tell when an adult is lying to you! Christiana heard the lie in the tone of her mother's voice; however, she called anyway to let her mother explain what she had called her about.

Once Christiana placed the call to her parents, Bash answered the phone. He usually asked his daughter how she was doing, but that time, he only said, "Hello," and then hurried, saying, "Here's your mother!" Bash tossed that phone off to Natalie like it was a hot potato! Christiana thought to herself, *well damn!*

Natalie got on the phone with an exaggerated sounding tone, stating what the supposed caller had said to her! Christiana said, "OK! Let me ask you a question. Did you all call the police since you received a call like that?" The next question was, "Why would you all be waiting for one of us to call you back if you did not reach us the first time?"

Christiana then asked, "How come you all didn't come to see if we were alright?" Natalie began to make excuses, stating, "That was why we called!" Christiana said, "We all could have been laying up in here dead, or we could have walked out into a line of fire because I know if I had received a call like that, I would have tried to move heaven and hell for mine!" Christiana realized that they thought she was not smart enough to understand how threatening someone's life causes a reaction that would make you run towards danger when your family is involved!

Niles and Cassius heard their mother talking and came to see what was going on! Niles took the phone away from his mother's hand. He then asked his grandmother about what was supposedly said on the call as that story began to change!

What neither grandparent had anticipated were questions regarding the supposed phone call! The story went from the caller being a male to a female caller, and it also went from stealing to someone being jumped. Niles asked, "Which was it? Because I'm confused. Was it a man, or was it a woman?"

Cassius took the phone from "Niles"; he was also asking questions, but there was another thing that "Natalie and Bash" had not anticipated, which was for their grandson's cell phones to start ringing! The other family members who were drawn into the shenanigans were the ones who rang their cell phones.

Meanwhile, Cassius asked to speak with his grandfather again, but that's when he heard his grandmother on her cell phone in the background talking to Nadia. She told her how they were questioning them! That was when Cassius asked Bash, "Who are the people that you claim you hear things from?" Cassius then asked him, "Why would you tell me not to tell my mother about what you said you been hearing from the last time that I was over there?"

Niles took the phone back from his brother and began asking more questions. That was when Bash gave the phone back to Natalie! Niles said, "I am trying to understand how someone would know who you all are and how someone would get your phone number?" Bash and Natalie were stressed once they understood that they could not control the narrative!

That was when Natalie said, "It's not like we aren't listed in the phone book!" Christiana heard what she said, which led her to look in the phone book and saw that they were not listed! Right then, Christiana got her cell phone and called 411 to see if they were listed, but again, they were not listed. Christiana said to her sons, "She's lying!" Christiana always taught her sons to listen closely to what a person is saying because people will tell you everything that you need to know.

Cassius started calling family members in no order, which caused them all not to be able to get their stories together! Once Cassius spoke to Lynsey, he began to ask her questions because her boyfriend had called him! It turned out that Lynsey was the one who ended up telling the truth about who the original story came from, which was from "Nadia and Natalie"!

Cassius called his grandfather back and told him that he knew who his sources were, which happened to be Nadia and Natalie. Bash was about to hit the ceiling, claiming that it was not Nadia or Natalie who concocted the story, and that was when Cassius told Bash that Lynsey had just told him the truth! Natalie was angry, and Cassius heard her in the background, stating that she wished she had never called them. Cassius asked if it was not true, why she would have wished she had never called.

Christiana could not believe that her family had done such a wicked thing after knowing what they all had gone through after a life-or-death ordeal and knowing that the loss of any of her children would have traumatized her. Natalie saying how someone was going to kill her son did not sit well with her daughter, and in that moment, Christiana lost all respect for her parents.

Chapter 65: That's On Me

The Most High asks Christiana a question shortly after that call from her mother. He asked her, "Why you are so loyal to these people!" The Most High asked if it was because they are your blood family. Christiana was looking in the mirror at herself and knew that she could not lie to herself! Christiana realized that it was because they were a blood family. The Most High then said, "That's amazing because you are not that loyal to me!"

Christiana felt like the blinders had been lifted off her eyes to the truth, and after that, her blood family appeared very different to her! Christiana was angry, and she realized how lethal her tongue was because of it, but she understood she did not want to do anything that would not be pleasing in the Most High's sight!

Christiana had a close family member who died, but she could not even bring herself to attend that funeral because she knew that her family would be there, and seeing them would have been all that it would have taken to set her off! Once she understood that just one fake hug or a venomous kiss on her cheek would have caused her to act out! Christiana also knew how much they thought of her and her children was too much for her to handle, and the best thing she could have done for herself was to distance herself from them until she was no longer lethal.

Christiana felt like after all the madness she had endured in her life, the only people she believed she could depend on were at least her sons!

Christiana had always told her sons, "Your best friend is your mother, but it also depends on who your mother is!" Christiana always told her sons, "You need to be careful how you treat your children because you will need them one day, and that the parent can't be mad if the child or children don't show up for them."

Christiana also told her sons not to be mad with their father because of his behavior towards them just, even if he had hurt them deeply. However, she told them to thank him for showing them the man that they should never want to be! What she did tell them was they needed to show their father how to treat them. The sons were also told that, hopefully, it would change him to be a better father; however, if not for them, for another woman, he would learn to treat that woman properly because they were already better men for how they treated the women in their lives!

Christiana always told her sons never to walk in anyone's shadows but to be their own man! The boys were also told not to have children until they had experienced some of life, and once they were able to buy their own toilet paper to wipe their butts along with their own soap to be able to wash their own butts, they should wait! Christiana explained to them that once they conceive any children, they should be prepared and understand that their lives are now lived for them!

Christiana made sure that her sons understood that once they married and had children. They should learn to leave a legacy for their family and that it's by how they lived their lives, which will be an example for them to follow because it teaches them who they are! Their mother explained that it would give them a blueprint to follow, but mostly, they needed to understand that parenting does not come with a manual; you just give your all because children are gifts from the Most High.

Chapter 66: Choose You Always

Christiana remained married to Tony for a while longer, but once his mother died, she had a clear understanding of the tree he had come from. She realized that if she remained married to him, it was not going to get any better! Christiana understood that the entire family had some mental issues even more complex than hers, and she was done with all the madness of fighting.

There was a heated argument between them right after his mother died over something simple and petty! Tony told her that things were going to be his way or the highway, and Christiana told him, "Trust me, it will be the highway," because she was not willing to be treated in a way that all she knew was not healthily minded! Tony went as far as to say to her that he was expecting repayments from her. Christiana looked at him and then stated that she wanted to see him collect it.

Christiana had done a lot of fantastic things for Tony, like making a house into a home, which saved them a lot of money. And once people visited the home, they always asked him who did the work on the house, and he always said, Christiana! One thing about it was that she exposed him to a new lifestyle that also involved adventure, along with several nice vacations that he could not have imagined because he had never experienced it with anyone.

Christiana made sure he knew how much she appreciated him for the things that he had done for her and her sons, but jealousy from his family was what he allowed to cause a rip further in an already shaky marriage! Tony allowed his family to get into his head, mainly listening to his envious father. Christiana recognized that he was going to target Tony first because his father had an unhealthy marriage with his wife until her death. Christiana told him that now that his mother was gone, his father would focus on him to destroy his marriage by using an allegiance tactic, and that's what Tony allowed his fathers to do like a stool pigeon student!

Tony knew that Christiana had brought value to his life and had changed for the better after he met her, but the one thing that he was not strong enough to be was his own man! Christiana was tired of being forced to wear pants and having to be the strength for everyone because only shelling out money was the easy part. One day, she thought that she needed to do a ball check because he was acting more like a female, and the order in which a marriage should be was far from what it should have been.

Tony chose to follow the poison that was being whispered in his ears regarding his marriage, but that was a deal breaker for Christiana. She realized that he was unfaithful even though it was not from another woman, but he was faithful to his father over his wife! Christiana understood that she was again unequally yoked with another man, and no matter how she provided, they would choose chaos over happiness!

One day, Tony asked her what she was waiting for to file for a divorce. She took her time listening, and then she told him that it was coming, and that's when he hollered, "When?" Christiana looked at him and then told him that he was a foolish man, but it would be coming sooner than he realized. Tony always spoke without thinking, not realizing that

Christiana was a woman who spoke literally; he had forgotten that she had told him that she meant what she said and said what she meant so that there would not be any misunderstandings! Christiana told him that she would not be the one to ever use the divorce card until she was given reasons to and that she was about to live up to all the ugly things he claimed she was not!

Tony had the egotistical nerve to tell Christiana that he needed her to go with him to see his lawyer, but that was when Christiana chuckled and then said, "I will be going to obtain my own lawyer, and I just look like this!"

Two weeks later, Christiana moved out and left everything at the house because, again, she was not interested in any reminders of him either! Tony was served with divorce documents, and apparently, after talking with his lawyer, he realized what he was standing to lose, which was more than he had bargained for, and that was when he wanted to humble himself, then called Christiana!

Tony wanted to know if they really needed to go through with the divorce. She told him, "Oh yes, sir, we need to go through with it because you campaigned hard for this. Besides that, I did not sign up for the madness or the level of disrespect!" Christiana let him know that once disrespect was served to her, it was too late to put it on rewind because she had never been nasty or disrespected to him!

Meanwhile, Christiana began to put her life back together and then began making life-changing decisions for herself and Ishmael because he had plans of going off to college. Christiana understood that she had poured a lot into her other sons. However, it was now "Ishmael's" turn for her to pour into. Christiana knew that Ishmael had not had a normal childhood because he was the one who helped carry the weight along with his mother!

Cassius and Niles were both of legal age by then and were living their own lives; however, they were no longer spending very much time with their mother or "Ishmael," and when it came time to spend time with their own family! The older sons seemed to choose spending time with other people rather than with those who valued them the most.

Once Christiana understood that her older sons no longer valued her or "Ishmael," she never said anything to them about it or how they treated them as if they were less than. That's when "Christiana and Ishmael" began to move accordingly due to their behavior. However, it made for an easy decision when it came time for them to move out of state.

Chapter 67: No Rhyme Or Reason

Ishmael called his father Di'Ban and asked him for twenty dollars for some lunch money one day because he saw that his mother was struggling while paying attorney fees. Meanwhile, his father told him that he would give it to him and never made any attempt to provide it for him.

Ishmael could not believe that he had left him hanging, and that was where their breakdown began! Ishmael became disheartened with his father and said it was selfish of him not to care if he was even eating; however, his mother made sure that Ishmael had money to eat by providing him with more than just twenty dollars.

Ishmael was involved in a serious car accident, and when Christiana called Di'Ban, which was something that she seldom did, but only if it were serious enough regarding his sons. Once he answered his phone, she began explaining Ishmael's unfortunate dilemma. Di'Ban immediately said, "What in the fuck are you telling me far!" Christiana hung up the phone because she still understood that he was who she knew him to be. Christiana knows that the day will eventually come when he's going to need Ishmael!

Ironically, Christiana received her finalized divorce decree on her and Tony's wedding anniversary. She had been planning for her new journey while waiting for the finalized decree. Christiana did the footwork on locations for her and Ishmael's move, and she also sought out schools for him. She was putting everything she had into their new journey.

Christiana Called and asked her nephew Manius if he would join her in taking a trip because she wanted to check out the area where she was looking to relocate to in Los Angeles, California. Mother and son wanted to be near the coast, where the weather was warm throughout the year.

Manius agreed that he would go with his aunt for support, but what was so nice about the trip was that her nephew had never flown before, and he had never seen the ocean either! Manius did not realize that it gave Christiana great joy to be able to treat her nephew with an experience that he would never forget!

The morning Manius and his aunt were getting ready to fly out to Los Angeles, it was cold in Chicago with temperatures in the single digits, and there was snow on the ground. Christiana told her nephew not to overdress because they were taking a straight flight, and the temperatures were going to be in the mid-eighties once they landed.

Manius said he could not believe what it felt like as they were loading the aircraft, and everyone was seated. His eyes were very wide open, but as the plane raised its engines, it began taxing down the runway for takeoff! Christiana looked at Manius, and then, in a funny tone, she said, "Just enjoy the ride, sir. We'll be there soon."

While they were in flight, Manius mentioned how he could literally see the change of seasons from the air as they moved across states. Christiana said, "Wait until we land. You are about to actually see the changes, and it is beautiful."

Their flight landed at LAX airport, and when they emerged from the airport, Manius could not believe how sunny and warm it was! They made their way to the hotel; after that, they went and got a rental car. Manius was flabbergasted by seeing the hills as well as the ocean and how they complimented each other.

Later that afternoon, they found their way to the ocean, and Manius could not believe his eyes. He kept saying, "We are sitting outside with shorts and flip-flops on in January," and he was astonished to see how the ocean met the sky!

Manius and Christiana found a spot at an outdoor café where they had a bite and some drinks while relaxing. That's when they saw all sorts of high-end cars back to back! Manius stated he had only seen these cars either in magazines or on television. Everywhere they went, he kept seeing more nice cars. He called one of his friends to tell him about his adventure while in LA and how unbelievable it was. The following day, they met with a realtor to find a home suitable for Christiana and Ishmael; however, it was going to take more than a week to locate a home that was perfect for them and within the price range. Christiana had already had another option in mind just in case she could not make up her mind.

Christiana had already done the research on apartments in the area that she could consider if everything worked out to where she wanted to stay. The day before they were to leave and arrive back in Chicago, she had gone and signed a lease because Christiana was sold on the move!

After the lease was signed, they went to dinner to reflect on all that they had enjoyed. Next, they stopped and purchased a bottle of Dom

Perignon. They made their way to the beach to celebrate the new beginnings that "Christiana and Ishmael" were about to embark upon. They were very excited but realized that they had to leave, but she told Manius that he could come to visit them at any time! Christiana also let her sons know that they would be moving to LA and that she had already located a resident, and unbelievably, her sons came to her asking if she was going to cook for a holiday. Christiana told her sons that she had not planned on it because they had stopped coming to partake in holiday dinners with her. The older sons told her that they would buy what she needed to cook and that they would come to help with the cooking.

Chapter 68: Disgustingly Good

After Christiana arrived back in Chicago, she had begun packing and putting her plans into action. The first thing that she did was go visit her parents to let them know that she and Ishmael were about to leave Chicago and move to Los Angeles within two weeks!

Once she arrived, Christiana informed them of the scheduled move, and Natalie looked like someone had just served her a straight shot of alcohol without any chaser! Christiana ignored the facial expressions because she had already expected that they would not be happy or would celebrate their new journey.

Christiana went on to tell them that the move would give them both a warm place to come visit during the cold winter months, but before that, she could not hear any sort of excitement from either of her parents regarding visits! All she heard were excuses of why they would not be coming. Christiana decided right then that she would not make another extended offer again.

Natalie instead chose to pick an argument, but Christiana wasn't for any arguing; she told her parents that the family was already broken because of all the lies and betrayal. However, it was repairable! Their daughter stated that all everyone needs to do is peel down their layers of lies to repair the damage that's been done.

Christiana went on to say that it's sad when you feel like you can't trust your own family, but it's even sadder when you feel you can trust a stranger first! Christiana told them that the reason she was not able to attend their family members' funerals was because of her anger towards them; she also said that the Most High told us all to "Honor thy mother and father so that thy days may be long!"

Natalie quickly jumped to say, "That's right!" Christiana said, "But hold on because if you keep reading, it also states parents do not provoke your children to anger!" Bash looked like he understood what his daughter was trying to convey, and his daughter saw that it spoke volumes to him but not to Natalie! Christiana said, "Listen, I am coming as humble as I know how, but if it does not matter to anyone else, then what I'm saying here is pretty much useless!" Christiana lived the majority of her life dealing with narcissistic people, which taught her how those people don't change because of their egos and prideful nature; however, what they will do is taint your character in order to smooth over their own misbehaving ways.

Christiana had also informed both Cassius and Niles from the beginning of her divorce that she and Ishmael would be moving out of state. She also asked them if they would care to make a move with them, but they declined and stated that they were staying in Chicago.

Once the two-week window had arrived, it was packing time to load the truck. Christiana asked her son Niles if he would be willing to drive her belongings to LA for her. Niles said that he would with one of his friends, and his mother gave him money to drive the belongings with a car carrier attached to the truck because it was too much for her to handle by herself. Apparently, the reality of their mother and brother leaving to move out of state set in on both her sons.

Niles was helping to load the truck for a while, but then he suddenly stopped and left! Niles returned early the following morning, just in time to drive his mother and his brother to the airport. Cassius did not attempt to show up to assist with any packing to help his mother and brother, nor did he even come or call to bid them farewell, which his mother felt was disheartening.

The one thing that Christiana had hoped for was that her sons would carry with them was how she had taught them to survive! Their mother wanted her sons to broaden their minds and for them to understand that there's more to life than what they could possibly imagine! Christiana always told her sons to live their lives to the fullest because there are no do-overs and you only get one chance at life, but mostly not to live it without any regrets, so choose wisely!

After Niles delivered all the belongings, his mother flew him and his friend, who rode with him, back to Chicago. After a few days, every time his mother looked up, Niles was coming back to visit. Christiana kept asking Cassius when or if he was ever coming for a visit to be able to enjoy something different. His mother went as far as to ask if she could at least have all her sons come together to visit at least twice a year. Cassius always came with excuses, but after being asked for two years straight, his mother and brother both decided to stop asking him. Cassius called his mother one day and told her that their father was going to come to visit "Ishmael." that's when she told him that he needed to talk with his brother about it, but they needed to find them a hotel and that Ishmael would come to them where they were! Every once in a while, Cassius would tell his mother how his father would tell him how he dreams about his mother and that he should not have ever let her go. His mother told him, "Well, those dreams were nightmares, and he did not have a choose as far as keeping her!"

Cassius was dating a female named Reagan, who felt the need to gain control over Cassius, and she did so by claiming that Christiana had an issue with her! The truth of the matter was that his mother didn't like that she would tell Cassius an entirely different version of conversations that never even took place. There had never been one time that Reagan actually set down to have a conversation with his mother to get to know her or to clear up whatever mishaps that she claimed took place because she realized that his mother could see through her and would call it like it was! Cassius apparently chose to believe what Reagan told him until, one day, he learned that it was all about her conquering and dividing for her personal gain!

Ishmael always told his mother that he knew that without her, he probably would not have ever been able to see the things that his mother exposed him to or to be able to experience! Ishmael said he wished that his brothers would have made the change along with them, but his mother told him that she hoped that because of their change, they would get out of Chicago to experience life elsewhere, but it needed to be their own dreams of desire.

Niles decided to further his education, but he needed a place to live after he and his girlfriend had a breakup; he then moved out from living with her until after he was to graduate! Niles first went to his father, Di'Ban, and asked him to stay with him. But his father told him that his girlfriend said that Niles was not welcome to stay with them out of pure selfishness! It hurt Niles that his father went along with it because he put his girlfriend over his own child, which was what he told his mother!

Niles then decided to go and stay with Tony until he graduated. However, Tony also had motives for letting her son stay with him; he figured that it might cause Christiana to reconsider being with him, and she would come back to him. Niles knew that his mother was never returning other than to visit her sons. Niles had considered moving to Los Angeles, and his mother told him that he would probably have great success there because the field that he was in was big business in California.

Niles called his mother to let her know that he was coming to visit for a few days but that he also had something else to tell her. Niles knew that she was not going to be happy about it, but he went on and told his mother that Tony was coming along with him. Christiana was hot and then asked, "Why and what for?" Niles said, "He offered to buy the airline tickets and had already purchased them because he said he wanted to see you and Ishmael!" Christiana said, "Son, you are going to owe me for that shit" Niles laughed then said, "Ok!"

Christiana had to mentally prepare herself for the visit. She took off the day that they were to arrive, knowing that jealousy would be arriving as Tony's personal guest. Christiana not only made sure she looked nice, but she also cleaned the house and prepared it very nicely. Then, shortly before they arrived, she had the new car detailed before picking them up from the airport.

Christiana needed Tony to get all his jealousy out at one time, so she thought!

Once she arrived at the airport, she saw "Niles and Tony" were waiting curbside as she pulled up! Christiana popped the trunk open, rolled the window down, and told them to hurry to place their luggage in the trunk because it was a fast-paced airport where they needed to move quickly.

Niles jumped into the back seat of the car, and Tony got into the front passenger seat. Christiana told them that it was good to see them. Niles smiled big but already knew that his mother was not really happy to see Tony!

Niles told his mother that he liked her new vehicle. However, Tony had no comment regarding the vehicle other than to tell her how nice she looked. She thanked them for the compliments, then popped in some riding music to keep the vibes light!

Once they arrived at her home, she had them wait at the front door while she put away her massive Cane Corso dog named "Smokey" that she had gotten for protection before letting them enter. Once they were settled in with their luggage, she had one more stop to make, and that was to show them the house that she was working on purchasing, which came with an inground pool!

After looking at the house, they went and picked up Ishmael from school. Then, they headed to the Santa Monica Pier to eat and enjoy the views! It seemed that everything was going better than Christiana had expected, at least until after they arrived back at the house.

Niles and Ishmael went for a walk, and that was when Tony came and said, "You did not even give me a hug when we first got here!" Christiana took in a deep breath and then said, "Listen, it was not a hugging time when I pulled up; it was actually a jump-and-go moment, and besides that, I didn't even get to hug my son either, so what's the problem?"

Right then, her sons arrived back in from the walk, and that's when Tony hurried up and changed the subject, not realizing that they were the least of his concerns! Now Smokey, her Cane Corse, always stayed near Christiana's side; however, he never took his attention away from Tony, but he was going to see how protective Smokey was when it came to Christiana.

The following day, Niles told his mother that their cousin on his father's side had contacted him that he was in LA and that Jamel wanted to see them because it had been years since they had seen one another. Christiana said it would be nice to see him also, after so many years, but that she probably had forgotten how he looked.

They all made their way to spend time with Jamel that afternoon, and once Jamel saw Christiana, he snuck up on her; he knew she would not recognize him! Jamel walked up and then hugged her tight. They sat, but Jamel kept looking over at her, which caused excitement from seeing each other again. They just hugged and squeezed each other while laughing! It had been so long since he had seen her, and they never thought that they would be seeing one another so far away from home!

Later that evening, after sharing the dinner with Jamel, everyone made their way back to the house. Niles and Ishmael ventured off to themselves away from the house, leaving Christiana with Tony. Tony then made his way into the kitchen, where Christiana went to pour them some wine. Meanwhile, he could not hold in his jealousy a minute longer when he said, "I can't believe that you took me to see another guy!" Christiana looked at him like he was out of his mind. At that point, she stopped pouring the wine!

Christiana asked him, "You do realize that we are not married, let alone a couple anymore, right?"

Christiana also said, "And it's not that I should need to explain that to you, but…" That's when Tony raised his voice to argue! Smokey stood up from the floor, squaring up, waiting for him to raise his voice at her again! Tony still carried on with a raised voice, but by then, Smokey stepped in front of Christiana with the hair on the back of his neck raised. He began barking, which echoed throughout the house! Christiana said, "I would advise you to bring your damn voice down because if he grabs your ass, there won't be anything that I can do to save you."

Tony realized that Smokey did not play when it came to Christiana. He backed out of the kitchen slowly and retreated to the living room out of Smokey's eyesight!

Christiana gave her dog a treat, picked up her glass of wine, and then went to her bedroom as Smokey followed behind her. He lay down at the entry of her bedroom doorway as Ishmael had taught him to do regarding protecting his mother.

Niles and Ishmael returned shortly afterwards, and when they came back in, Tony was looking shaken, which they had seen on his face. They then asked him, "What is going on?" Tony was telling them how Smokey had gotten aggressive with him for just talking with Christiana! They both knew that something more had to happen for the dog to have acted in an aggressive manner because that was how he was trained.

Ishmael told them, "I'll be right back!" Niles understood when he saw that his brother went straight to his mother's bedroom with the corners of his mouth curled up! Ishmael rubbed Smokey on the top of his head and then told him, "Good boy!" Ishmael was snickering when he asked, "So what did he do?" Christiana said, "He thought he wanted to raise his voice until Smokey put that ass in timeout. Basically, he told him to get his ass on that couch, and he bet not move!"

Christiana asked, "I bet he was on the couch when you all came back in, wasn't he?" Ishmael said, "He was looking silly in the face, acting like Smokey had gotten aggressive with him for no reason, but we knew better! I just came to tell him good boy and to find out what he had done!"

A couple of days later, Christiana told everyone the night before that they needed to be packed and ready for their flights. Once they picked her up from work, she could get them to the airport in time. Christiana comes out of her job expecting to see everyone in the car and ready to go, but when she comes out, she only sees Tony!

Christiana did not like the fact that he was driving her vehicle, and when she got into her car, she asked, "Where was everyone?" Tony said, "I told them that I would pick you up," and Christiana said, "I specifically said for everyone to come together so that we could get in rush hour traffic in order for you all to get to the airport, so you both could make your flight on time!"

Once they arrived back to pick up Niles and Ishmael, her sons could see that Christiana was pissed. All she said when she walked in was, "Let's go!" Everyone grabbed their bags. Christiana then said, "I am in the car." Ishmael looked at his mother as he had a hard time containing his laughter because he saw that she had had enough of Tony and was ready for him to leave to regain the peace in the home.

They all jumped onto the crowded freeway in rush hour traffic. Christiana was not saying anything because she was focused on jumping around in traffic and trying to get them to the airport in time! Christiana realized it was cutting it close, and that was when Niles said, "We are probably not going to make it in time." Christiana said, "That's why I said for everyone to come when I got off of work," and Niles looked at Tony to let his mother know it was all on him.

Christiana responded and said, "Well, I'm trying anyway!" Tony said, "We could probably stay a couple of more days." That's when she said, "I'll get you all there. Now, you all might have to run to make it onto the plane, but I'm trying!" Ishmael sat back in the seat, smiling at Niles, letting him know she meant what she said.

Christiana made it to the airport, dropped them off, and headed for the freeway. Ishmael said, "You were mad." Christiana said, "I couldn't take another day of it." Christiana's cell phone rang, and Ishmael answered it! Ishmael said, "Niles said that they got there too late and that they needed to come back to pick them back up." Ishmael said, "He intentionally told them not to ride with him to come to pick you up while he was trying to stay longer!"

Christiana was livid. She turned the car around to go pick them back up, and once they got in the car, Niles said, "We can make an early morning 6 am flight because he knew his mother was fed up after that stunt!" Tony again said, "We can stay a couple more days, and I could extend the flights and pay the fees for it." Christiana did not ponder and said, "No, I'll get you all there for that flight on time first thing in the morning at 5 am; therefore, you all won't miss that flight!"

Tony stated that he wanted to take everyone to dinner, but "Christiana's" mind was not going to change regarding getting them to the airport the next morning based on dinner. Christiana told them to hurry up and eat because she needed to get back home to get ready for the next day. Once they arrived back home, Christiana retreated to her bedroom and went to bed early to make sure she got them there on schedule, and nothing was going to stop that.

Los Angeles was under heavy rain throughout that night, which woke Christiana from her sleep around 2 am! As soon as her eyes opened, she set straight up from her sleep without haste; she slid her feet into her shoes, which were sitting beside her bed! Christiana grabbed her keys and purse from the dresser; she then cut on all the lights in the house and told them, "Come on, let's go!" They all looked stunned by how she cut the lights on, blinding them. She said, "I'll be in the car."

Christiana then walked out, waiting for them to come. She was determined not to allow anything to slow her down. She did not even bother to put on clothes and left with only pajamas on! Christiana dropped them off so quickly, then wished them well and peeled off before they had the opportunity to enter the building.

On the ride back, Christiana finally realized how hard it was raining. She said to herself, that's how you can tell when you are sick of someone when you don't care what the weather is doing. All you knew was that he had to go! Once she arrived back home, she went right back to bed and slept well.

Later that evening, Niles called his mother, and when she answered, he began laughing. He said, "You know what?" Christiana was laughing when he said, "You could have at least put some clothes on!" Christiana said, "I didn't have time. I could not take another minute of him!" Niles was still laughing when he said, "It was 2 am, and our flight didn't leave until 6 am. I saw what time it was by the way you cut the lights on. I was like her and "Smokey," both sick of his ass, then I looked back to wave byc to you, but you had cut out. All I saw were tail lights!"

Tony's nephew "Felipe" called to ask if he could come to Los Angeles to live with them, and she told him that he could because he and Ishmael had grown up close to one another. Once he arrived, both the young men were enjoying the stay, and it was through phone conversations with his nephew that Tony decided that he wanted to come for another visit.

Christiana told him that he needed to get a hotel room for that stay because she did not have any room for him to stay with them; she did not want to deal with his foolishness in her home anymore.

Christiana had booked a hotel stay for once he was to arrive. She figured that they all could go to Disney Land with the younger men to let everyone enjoy that activity. When he arrived, she and the young men picked him up from the airport, and he instantly stated that he didn't mind sleeping on her couch. She instantly told him that her couch was not for sleeping on.

Tony was then taken to his hotel, and they sat with him for about an hour. They told him that they would see him the following morning. The next morning, Christiana woke up the young men so they could all get ready to pick up Tony and head to the theme park. Ishmael and Felipe decided that they did not want to go with him to the theme park, which she was not happy about, but they figured out what the worst could happen! Christiana picked Tony up from the hotel but was hoping that it would be a healthy outing.

Tony came down from his hotel room and then got into the car. He apparently did not like the fact that, in his head, he felt she was too much in control! Tony decided he was going to insult her and try to knock her off her square because she was unable to look at something, and he kept trying to show her on his phone while she was driving. Christiana told him that she couldn't look at the phone because she needed to keep her eyes on the busy freeway. That's when Tony said, "With your fat ass!"

Christiana slowly glanced over at Tony while giving him a once-over from head to toe. She chuckled and then said, "As if your ass is a picture of perfect health!" Christiana hit the next exit off the freeway and then said, "I tell you what! I'm about to drop you off at that hotel so that you can take and get on the first thing that you can catch because you keep forgetting that we are not in any sort of relationship, and disrespect was part of the reason as to why not!"

Tony called Ishmael and Felipe, pleading and explaining that he said some things that he knows he should not have said to Christiana and that he wanted to apologize! That was when they told him that she was not interested in anything that he had to say. They also told him that they would take him to the airport to see him off and for him to let them know when he was ready.

Felipe's father, Felipe Sr., who is Tony's brother, called Christiana to talk with her about his brother; he immediately stated that he had talked with Tony and told his brother that he did not understand how he messed up as he had and by letting their father get his head concerning her.

Felipe Sr. asked if Tony were to sincerely apologize, would she reconsider taking his brother back? Christiana told him that she was not interested in his brother and that she was off that ride! Christiana did say the best thing that he could do for his brother would be to get him some clinical mental health help because he needs it!

Chapter 69: When Duty Calls

Cassius called his mother on a daily basis just to talk with her and gain wisdom. On that particular day, once he called her, she heard something else in his voice! Cassius told his mother that he had received a call from his transplant team and that they told him that he needed to stop everything that he was doing and come to the hospital because he was in rejection.

Christiana dropped what she was doing and then told Cassius that she was on her way! Christiana grabbed her a bag, and then she had Ishmael take her to the airport. She purchased a straight flight to get to her son soon. Christiana finally arrived in the city late in the evening; she then headed straight to the hospital because she understood she needed to jump into her assigned coaching mood for her son.

Christiana was moving fast to get to Cassius's hospital room, and once she arrived, she could see that he was fearful of what was about to happen! Reagan, his girlfriend, was there with him when his mother walked in after she spoke to her son. Christiana went out into the hall to speak with the nursing staff. Reagan followed her, and then she told Christiana that she could not wait for her to get there because she didn't know what to do!

After speaking with the nursing staff, Christiana jumped into her motivational role, which she was accustomed to, trying to calm her son and reassuring him that everything was going to be alright!

Christiana stayed around the clock with her son for a week and a half, and once he was released, she was going to stay in a hotel for the remaining two days before she was to leave to go back to LA! Niles asked his mother to come and stay with him at Tony's house so that he could spend time with her. Christiana told her son that she was not interested in staying there, but he assured her that he would be there with her for the remaining length of her stay.

Against Christiana's best judgment in trying to please Niles, she went and stayed at the house with them. Tony was told that she was coming to stay for a couple of days, and once Tony came in from work late that evening, Christiana thanked him for letting her stay there until she would be leaving! Niles ended up going out each night, leaving his mother alone with Tony!

Somewhere around 1 am, Tony asked Christiana where she was sleeping for the night. She told him, "In one of the other two bedrooms," where she would make her a pallet! Tony thought she was going to sleep with him; however, Christiana had no interest in lying next to him.

Christiana had been a structured person for a long time. She's one who always puts her things of importance in either one or two places! Christiana had placed her keys, her glasses, and her purse together on an end table in the living room.

The next day, when Christiana arose from her sleep, she had gotten on her laptop and was minding her business quietly in the living room. Tony arose and then went to where she was. Christiana looked up when

Tony walked into the living room; she spoke pleasantly and then went right back to what she was doing.

Christiana remained on her laptop until after Tony had left to go to work, but apparently, he had a problem with the fact that Christiana was not giving him the attention that he must have desired.

Once Tony arrived back later that evening, again Christiana was on her laptop, and he came to where she was as he was making conversation with her. Christiana was conversing with him; however, she did not set her laptop aside while they conversed. Again, she set her belongings together on an end table before she went to rest her body for the evening.

The next day, Christiana arose and did not go straight to her laptop; that time, she decided to watch television instead. Tony did not come to make very much conversation with her; he stayed in his room until he left to go to work. Once he left, Christiana decided to get her laptop; she went to where she had left her glasses, but they were not present. Christiana went into the bedroom where she had slept to look for them, but they were not there either.

Christiana remembered exactly where she had put her belongings. She immediately retraced her step from the night before; Christiana was already thinking, *I bet this silly fool has taken my glasses* because she was not looking at him and knowing that he felt like she would be looking at nothing else either!

Christiana franticly looked through the entire house for her glasses. She looked in each room, the trash, the cabinets, the refrigerator, and the bathrooms, even in her son's room, but could not find them anywhere! Christiana called Tony. She asked him if he had seen her glasses, and his first response was, "So what do you think that I took the glasses?" Christiana said, "I know where I left my glasses, and I have searched this house looking for them, and that's why I'm asking!" The only thing Tony said was, "I didn't take your glasses," but Christiana heard the lie in his voice.

Niles arrived from school, and Christiana asked him if he had seen her glasses; he said that he saw them on the table from when he had arrived the night before! Christiana told her son she believes Tony has her glasses! Niles went to search his own room and looked throughout the house but found nothing. Christiana was livid about him taking her glasses. She said, "Who steals someone's glasses because they aren't looking at them? Now, how sick is that? I am not getting out of here fast enough!"

Christiana went and packed her bags to be ready to leave the next morning. Niles was going to take his mother to the airport and had set his clock so that they would be on time! Tony arrived back at the house late in the evening, and again, Christiana asked him if he was sure he had not seen her glasses.

Tony again became offensive, stating, "I didn't take your glasses." "Niles and Christiana watched how guilty he appeared! Niles and his mother watched television until late in the night, and once they went to bed, Tony waited until he knew that they both were asleep. He then crept into Niles's room and laid Christiana's glasses on Niles's bed! Once the alarm went off, Christiana heard it, and once Niles went to get out of his bed, he flung his cover back, and that was when he heard something hit the floor! Niles stated that he had almost stepped on her glasses.

Niles heard his mother coming out of the restroom. He met her in the hallway and then handed her the glasses. Niles said, "He had your glasses, then he snuck his ass in while I was asleep and laid them on my bed." Christiana said, "He needs some medications for sure after that."

Chapter 70: Set Apart

Christiana had returned home and was lying in her bed, resting the following day. Not thinking of anything in particular, her television was on, but the volume was turned all the way down. In the living room, Ishmael sat with earbuds in his ears, engrossed in music on the computer, creating a quiet atmosphere in the house.

A sudden thought crossed Christiana's mind: what color are the Hebrews? She lay there for a moment, contemplating the origin of this thought. The next thing she heard was a voice within, advising her to take it to the altar and leave it there.

Realizing that the "Most High" was speaking to her, Christiana rose from her bed and went to where Ishmael was. She asked him to Google, "What color are the Hebrews?" Ishmael did as his mother asked of him; he complied, turning around to look at her and responding, "They are black." Christiana's eyes widened, and she politely asked Ishmael to move so that she could further research.

Christiana comprehended that throughout her lifetime, the truth about the so-called African Americans, Latinos, or Native Americans being the true Hebrew Israelites had been concealed. The Most High was awakening Christiana to this truth, urging her to return to Him—the Father of Abraham, Isaac, and Jacob, also known as Israel (Isaiah 10:21-22).

For every question that crossed Christiana's mind, she sought to find discrepancies in the word. However, through the scriptures, she confirmed the truth. Christiana dedicated two weeks to deep study, even visiting the library for additional research. The blinders were lifted once more, allowing her to see life's truth according to the guidance of the Most High.

Christiana realized that the "Most High" chose her to be awake, and the reasons why those who had mistreated her were due to mass following because she had already been selected for being set apart! By the fourth year, these truths had become deeply embedded in the fiber of her existence as she continued her steady learning journey. Christiana was learning more on her journey about how the "Most High" was going to reveal even more to her! Christiana discovered those who were not equally yoked with her, and the "Most High" had shown her who they all were!

Unexpectedly, Christiana received a phone call from a family member informing her that her father, "Bash," had been rushed to the hospital. She was asked if anyone from her family had called her to let her know that it looked serious. Christiana said that no one had called to let her know anything; however, she appreciated her calling to inform her.

Christiana asked her to have someone call her to let her know what was going on with her father and that she was getting ready to pack her bag in case she needed to come right away. Unfortunately, throughout the remainder of the day, Christiana did not hear from anyone—neither her mother, Natalie, Nadia, nor Andre—to provide updates on her father's health. The next morning, Christiana, waking from her sleep, was dismayed that no one had contacted her. She picked up the phone and called the hospital directly to inquire about her father's condition.

Once she was transferred to "Bash's" hospital room, it was her niece who answered the phone. Christiana spoke to her pleasantly and then asked to talk with her father. Before the phone was given to him, she overheard the family whispering in the background to each other that it was "Christiana"!

Christiana asked her father, once he was on the phone, what was going on with him. He told her that he was not certain and that they were waiting for the doctor to inform them of more. That's when she told her father that she would like to know what was going on with his health. Later in the evening, Natalie finally called Christiana, and the first thing she began to say were excuses for not calling earlier. Christiana talked over the excuses because, at that point, it did not matter.

Natalie finally told Christiana that he was having some complications, which led them to take him to the hospital and that his test results showed some spots that they needed to biopsy. Christiana asked her mother to call her to let her know when that was going to take place.

Deciding to stay home the following day, expecting to hear from her mother about the procedure, Christiana never received the call. The next morning, Christiana again called the hospital and was able to reach her father, "Bash." When they began talking, Christiana asked him what was going on. He told her that they were scheduling him for a chest biopsy surgery today. Ishmael was listening to the call on speakerphone, and he went online to schedule an airline ticket for his mother, leaving within the next two hours.

Ishmael came back into the room with the luggage bag that she had packed, and then he told his mother to get ready because he had already booked her a flight. Christiana told her father that she was on her way. Within ten minutes, Natalie called Christiana, telling her that she did not need to come. Christiana did not say much, nor did she ask her any questions; she then got off the phone.

Christiana was standing in the middle of the floor in a daze, thinking to herself. Ishmael said, "I've got your bag because that doesn't even make any sense for her to say that you did not need to be there!" Ishmael said, "I've sent the flight information to your phone; that you'll need to board the flight, and I'll be in the car waiting!"

Christiana's flight landed, and it was Cassius and Niles who were waiting for their mother because Ishmael had called and told them what time her flight was going to land. Ishmael also texted his mother, letting her know that her other two sons would pick her up from the airport and then take her to the hospital.

Christiana made it to the hospital just as her father, Bash, had made it out of recovery. Once she entered his room, she saw her father; his eyes lit up as if he could not believe that she had made it. Christiana greeted everyone in the room when she entered. Natalie and Nadia looked displeased when they saw that Christiana had come anyway. Nadia gave Christiana a disdainful look, and Natalie's mouth literally dropped open. Christiana went to hug her father, but the tension escalated when Natalie asked, "Did you speak to Nadia?" Christiana looked at her mother and said, "I spoke when I walked into the room." She wanted to say more, but she understood she did not come for their foolishness. Nadia looked at her mother and said, "I am out of here; I'm going to the lobby." Christiana really wanted to say, "Well, bye," if you thought your presence made a difference because it doesn't!

Bash asked Christiana if she would shave him because his face was irritated. He immediately went on to say that the only other person who had ever shaved him was her grandfather, Ray Montgomery. Christiana gathered her father's shaving equipment from his bag, and she also got a tub of warm water and towels to shave her father. Natalie rushed and said that she already had someone coming to shave Bash, but Christiana ignored what her mother was saying because her father had asked her to shave him.

Once she was done shaving him, she took a seat beside his bed, sitting quietly and watching television with her father. Bash did not realize that Christiana noticed how he was looking at her while she was shaving him. Once she sat down, her father sneaked and slid his hand down to the side of the bed so Christiana could hold his hand. She sat and held his hand while they quietly watched a cowboy movie together, just like old times.

Bash realized at that moment just how they had unjustly failed their daughter. She also saw and understood that if anyone knew what he was feeling outside of the family, it was Christiana and Cassius. When Christiana and her sons were leaving the hospital, she told her sons that their grandfather had stage 4 cancer and that she would be surprised if he survived 30 days. The comment stunned her sons, but they knew she had seen sickness and death more than once.

Natalie never asked Christiana if she was even staying or how long she might be. The following day, Christiana called her mother, Natalie, to find out if they had received his diagnosis. Natalie said that it was in, but she told the hospital to wait until Nadia was available to be there to read it.

Christiana realized the nonsense that always comes with either of them, causing her not to call back because it was going to be more of the "yoyo" foolishness! Christiana had a return flight scheduled to leave in a couple of days. Meanwhile, Bash had been released to go home after she called to check on him, and that was how she found out.

Christiana called Natalie to check on her father the day she was to be leaving. She told her mother that she wanted to stop by before she left. Natalie said to her that he was not up for any company. Niles told his mother to forget what she had said, and he's taking her over there before she leaves because she isn't company; she is his daughter.

Once they arrived at the house again, Christiana was met at the front door with attitude from both Natalie and Nadia. Niles and Christiana came in and took a seat near Bash, who was up watching television. They engaged in lighthearted conversation, but Christiana could tell that he was not feeling well. Natalie and Nadia were trying to force him to eat; however, Christiana saw that what they were trying to feed him was not a good idea because it was some fried chicken and a bowl of heavy soup.

Bash had passed gas, and Natalie became mad about it. Bash looked afraid; he then picked up a bottle of some hospital air freshener and tried to spray it in the air, but it also got all over his food. Christiana got up from her seat, took the plate away, and said he couldn't eat this because he accidentally sprayed it over his food. The fact that Natalie had no compassion for her husband, once Christiana saw his reaction towards Natalie being upset because he had passed gas, was too much for her to handle.

Niles felt the same way because he made an excuse and then said, "Mommy, we need to get ready to go so I can get you to the airport on time." Once they got into the car, Niles said, the entire energy shifted in the room, and I knew you were not going to be able to hold it much longer.

The day after Christiana arrived back home, she received a call from one of her aunts, who was asking her how Bash was. She said that when she talked with Natalie, all Natalie was doing was talking about herself and how tired she was!

Christiana's aunt told her that when she talked to Natalie, her mother said to her that Christiana came but didn't do anything other than have a good time with her ex! Christiana told her aunt, "One thing is for sure: I didn't need to leave LA to go. Good timing, so I don't know what she's even talking about or what she expected from me!" Christiana called her father and found out that he had stage 4 cancer; however, it did not come as a surprise. She told her father that she understood, and she also understood the fear that he was feeling! Christiana immediately asked her father if he could do her a favor and fight because it's not what you go through but how you go through it! Bash began to cry because he was talking with someone who was understanding of what he was enduring and not in denial of the diagnosis.

"Christiana understood that he needed to be loved like never before, and he also needed to be able to talk about what he was feeling. Unfortunately, Natalie refused to let him open up about his feelings and fears. She disliked the fact that he became emotional and shared his diagnosis with Christiana. Christiana, who bore a striking resemblance to "Bash's" mother, "Madeleine," sometimes treated him as if she were his mother, similar to what "Gia" had done.

Regrettably, even in the darkest hours of a person's life, when they are on their sickbed, some individuals will act in ways that reveal their worst sides. Lynsey took to a social media site, making slick remarks about family for others to comment on. Lynsey lacked the courage to engage in direct communication but chose instead to engage in indirect actions, reminiscent of her habit of throwing rocks and then hiding her hands.

Christiana responded to Lynsey's post, stating her perspective without resorting to mudslinging. However, Lynsey's minister decided to share her sorry-ass opinion in the conversation. Christiana, wanting to address the issue directly, she sent a private message to the minister, explaining her relation to Lynsey and advising her to keep her comments to herself as she did not know the family beyond what she had been told. Christiana also told her that instead of entertaining some foolishness online, she should be more concerned with letting Lynsey know that it was not the time or place for any family drama! What she did tell her was that as a minister, her focus should be on bringing the family together in prayer instead of entertaining division.

Christiana ruffled the minister's feathers, setting her straight as the minister attempted to fire back at Christiana; she was about to learn a valuable lesson at the same time! The first lesson was that she was not such the holy roller that she pretended to be. She told Christiana that she didn't know her and nor did she care to know her!

Christiana did not allow that asinine comment to keep her from teaching the minister what she needed to practice, and that was when Christiana told her that it was perfectly fine if she didn't care to know her because she could be a lost sheep! What she did tell her was that the Most High would accept her wherever she was. Christiana realized the minister was too focused on defending her title and missing the mark each time she engaged in an argumentative debate. Christiana kept telling the minister that she was missing the mark, which infuriated her each time!

Christiana persistently tried calling her father each day, but Natalie consistently told her he was asleep. Natalie then made the mistake of revealing her intention to make Bash tell Christiana that he no longer wanted to talk to her. Undeterred, Christiana continued calling until the end of the week and got the same response; she waited to when she figured that he should be up by the next day. Again, Natalie said that he was asleep, but at that time, Christiana told her mother, "Well, when he wakes up, ask him to call me, please."

Natalie, following through on her earlier plan, called a family member to let them know her intentions. The family member who she had already told what she conjured up was a wicked plan regarding her daughter, which was about to happen! That same family member called

Christiana to let her know what her mother was getting ready to do; she stated that Natalie should be ashamed of herself. Christiana said, it is ok, I am already prepared for it; sadly, he is at her mercy in his last days, but he should have stood up to her long ago.

A few minutes later, as Christiana anticipated, her cell phone rang while she was at the car dealership getting an oil change. It will be a day that she will never forget! Natalie called with a hostile tone and then said, your father does not want to talk to you anymore! "Natalie's" venomous words fell upon deaf ears because Christiana already knew where it was coming from! Christiana heard her father crying his heart out in the background as Natalie demanded he take the phone and tell Christiana himself.

Bash was forced to take the phone, and once he heard Christiana's voice, he could hardly speak. Meanwhile, Natalie stood nearby, harshly saying, "Tell her!" Christiana said, "Calm down, Daddy, it's okay, really." Bash wanted Christiana to read between the lines; that's when he said, "I don't mean any harm." Christiana reassured him, saying, "Daddy, it's really alright, and I love you, so don't cry." She told her father, "What I need you to do is rest, and believe me, I already know where it's coming from, so rest!"

Once the call ended, Christiana realized that no one should ever know the day or time that it would be the last time they talked to their loved ones. Christiana also understood that Natalie did not comprehend that she had just helped expedite Bash's demise. Christiana thought her family could not get any lower, but she was about to find out just how low and dark their hearts would go to hurt her.

Christiana went home and shared with Ishmael what Natalie had made Bash say to her. You could see that he was having a hard time processing that his grandmother could do such an evil thing. Ishmael asked his mother if she was alright, and Christiana told him that she was. The only thing she realized was that she knew it was the last time she would hear her father's voice, and no one should know that.

Christiana told Ishmael and the other family members who called to check if she was all right after what Natalie had foolishly done. Everyone was surprised that it did not leave Christiana distraught and heartbroken. What she told each one of them was that her mother just helped to kill Bash much faster, and Natalie was more interested in creating foolishness, even in his last days.

Chapter 71: Funerals, Weddings, and Births

Christiana did most of her grieving for her father before he died. She prayed to the Most High to take him swiftly so that he might not suffer long. Christiana received another call from the same family member, informing her that the doctors said her father, Bash, had suffered a massive heart attack and was still having them. He was not expected to survive throughout the night.

Natalie, Nadia, nor Andre bothered to call Christiana to tell her that she needed to get there because Bash was not expected to survive. Instead, they took the time to reach their friends, associates, other family members, and pastors but not their daughter or sister.

Christiana was wrestling with the decision to come because she already knew what it was going to be like. She did not want her last memory of her father to be seeing him suffering on his deathbed. Christiana told Ishmael that his grandfather was not expected to survive the night. Ishmael also said that he did not want to see his grandfather in that state, but he would be going to his funeral services.

Christiana kept praying, asking the Most High for guidance. He led her to go on the third day before Bash was to transition. Christiana and Ishmael took a flight home to see her father before he drew his last breath. Once Christiana got to the hospital and walked into the room, as soon as Natalie saw her, she immediately pulled her face up and began crying, looking for sympathy but only for herself.

Christiana spoke to her mother, then went straight to her father's bedside. She began calling her father to let him know that she had arrived. She could see that he was in a semi-unconscious state, not being completely aware. Once Bash heard Christiana's voice, he was able to grunt to let her know that he knew she was present.

Christiana saw that her father's breathing was labored, and as soon as a doctor walked in to check on Bash, for some apparent reason, the doctor gave his full, undivided attention to Christiana. That was when she questioned the fact that it appeared like her father was laboring. The doctor listened to his heart and breathing, agreeing that he was laboring.

Christiana then asked how they knew if he was in any pain since he was not quite aware. The doctor stated that he probably was in pain and that they needed to give Bash more pain medication that would ease him from labor. The doctor also said that it would probably speed up the process of death, but it would make him more comfortable.

The doctor waited for a response, and that was when Christiana said, "I don't want my father to keep suffering." Natalie then said, "I don't want him to have anything because I am not ready for him to leave yet." Christiana damn near lost it; she then said, "It's not about none of us being ready for him to leave, but it is about him not being in any suffering pain." The only one that Natalie listened to came from Lynsey. Suddenly, she was agreeable to let the doctor decide to relieve him from laboring and increase his dosage of pain medication. Christiana went back to her father's bedside and took a seat, but as soon as she did, Natalie began with her shit.

Natalie asked Christiana, "Well, did he grip your hand?" because he gripped ours. However, Christiana did not respond because she couldn't understand how her mother could be so insensitive as to worry about whose hand Bash was gripping or not gripping. Christiana also understood that the gripping was due to nerve pain from people touching him, which happens during the dying process.

Later in the evening, when everyone was calling it a night, Christiana did not understand how they all were able to leave to go home and leave Bash all by himself. Christiana told them that she was staying with him. Around 3 am in the morning, Nadia and Lynsey returned, walking back into the room and turning on the lights. Christiana had just made a sleeping cot for herself to rest in the room along with her father.

Christiana got up, checked the vital machine, and then told Nadia that he was getting ready to die. Christiana then took a seat by her father. Nadia asked Christiana if she thought she should go to get their mother. Christiana looked at her and remembered how she would end up being blamed for all the crap that Nadia had created. Christiana said, "I don't know, that's on you." Nadia and Lynsey were so engrossed in a self-loathing conversation that they did not even realize Bash had taken his last breath. Lynsey eventually turned around and realized that he had died. She looked at Christiana as if she had control of Bash's moment of death. Christiana looked back at her and said, "Yeah, y'all missed it; he's gone," as they watched Christiana reach up and catch her father's last breath.

Nadia said, "I'm going to get our mother," and when she returned with Natalie, the first person Natalie looked at was Christiana. Other family members had already arrived before Natalie, but Christiana understood why she looked at her first before going to her husband's body.

Christiana's mother may have been able to control Bash when it came to him not being allowed to talk to her in his last days, but the 'Most High' allowed 'Bash' to wait for his daughter to get there and allowed Christiana to be with him until he drew his last breath.

Christiana went to her parent's house after they all left the hospital, planning to stay with her mother. However, Natalie asked Cassius as soon as he walked in if he had called to let Di'Ban, his father, know that Bash had passed. Christiana knew this was no more than a dagger that she wanted to throw at Christiana for being there with her father when he died.

Christiana thought about how she had Di'Ban on her mind when she never once had her own daughter on her mind to call her and let Christiana know that her father was not expected to survive. Christiana told her son Cassius that she needed to stay with him after that.

The ministers that Christiana had to chastise, her father came to the house, and not once did he ask to speak with the family in unity. Instead, he took Nadia and Lynsey into another room away from everyone to talk about whatever they needed to discuss in private. What they did not know was that Christiana did not care about it because she did not want to deal with any of the foolishness that she knew would come with this.

Once they came out of the room, the minister's father got to the door to leave. He turned towards the kitchen where Christiana and some more relatives were and then said, "Alright, homeboys!" Everyone looked at him like he was out of his mind and being so disrespectful to people that he did not even care to make his presence known, but it was only out of respect that those family members did not let his behind have it.

Rena, the next-door neighbor, came over to offer her condolences. She began telling Natalie that she remembered that when her husband died, when she arrived back home after making her deceased husband's arrangements, she found that her house had been cleaned. Rena asked who cleaned her house, and her family told her that it was Christiana who had come to clean for her so that she did not have to worry about it for her guests. Rena said how pleased she was that Christiana had thought about her. Natalie looked like someone had force-fed her a lemon.

Cassius was riding with his mother, taking her to his place as she rode quietly. Cassius asked his mother if she was alright because she was very quiet. Christiana told Cassius that she was okay. Cassius then said the family was devious about how they treated her, and he knows Christiana says that she's accustomed to how they act. However, Cassius went on to say that the low-down, spiteful behavior of his grandmother does not make sense.

Cassius suddenly stated that he realized how much Christiana was like his grandfather Bash, but that was of no surprise to Christiana. Her response to her son was, "Yeah, I know, but the difference between my father and me is that I am strong enough to walk away from the people and things that no longer serve me anything but good."

Christiana went on to say, "Son, just when you think that people can't get any lower, they will show you just how low they can get. What I do know is that there is nothing else that they can do that would shock or surprise me."

Prior to Bash passing Nadia, Andre did not say much to Christiana once they saw that she had arrived, but neither did Christiana because she was done with all the fake pleasantries. Once Andre had an audience, he began taking the position of leading the family. He said, "We need to bid our farewells to our father while he is still here, and I will go first. Then I will let everyone else who is going to speak with him next." This was no more than Andre and Nadia both were vying for the position prior to their father's death.

Christiana was then directed to speak to her father next. She told her father how great of a father he was and had been from every aspect of her life. Christiana said to Bash from the perspective of a young girl standing on his feet learning to dance, then remembering running to greet him from the bus stop every day and how he made each of those times feel like it was new each day. Christiana told him that everything she had learned, she had learned from him. Christiana wanted to thank him for being the father that he had been and let him know that she would miss him.

Christiana apologized to her father for anything that she may have done in life. Truthfully, Christiana only said that to ease her father's conscience, mind, and heart about the last spoken conversation they had so that he could go in peace.

Chapter 72: Opportunity for Insults

Natalie took the floor next, addressing Bash with endearing terms like "honey" and "baby." Christiana's eyebrows furrowed in confusion as she realized that, in her entire lifetime, she had never heard her mother refer to her father in any way other than Bash, and always with a hostile tone, as if scolding him.

Christiana glanced up at her mother, resisting the tip of her tongue to retort, "Now you can't even believe that shit for yourself." She saw through the act; it was merely a show for friends who were oblivious to the truth. Natalie seemed compelled to outshine Christiana, asserting that they did their best with the children after Christiana claimed to have learned everything from her father.

Tears welled up in Christiana's eyes as she left the room, seeking solace in the restroom to wipe away the evidence of her emotional turmoil. In the hallway, Nadia feigned concern, asking, "Are you okay?" Christiana, sick of their antics and the accompanying foolishness, restrained herself from expressing what she truly wanted to say, "Girl, fuck you!"

On the day of the wake, Christiana occupied a front-row seat reserved for immediate family. However, she preferred to sit far away from Natalie, Nadia, and Andre. Despite her wishes, Christiana's auntie insisted she move closer to them. Christiana said, "Please, no, I am fine with where I am," but she would not let Christiana be as she asked.

Christiana sensed the impending danger emanating from their presence, explaining her desire to distance herself. She felt her heart pounding hard and fast. Overwhelmed, Christiana wept uncontrollably, unable to look left or right, as if frozen in her seat. A well-timed comment or touch could have triggered a storm of emotions in the wake.

Once the last person spoke at the wake, Christiana hastily exited, craving fresh air. One of the relatives approached her, acknowledging that they had noticed her distress and expressing relief that Nadia didn't attempt to offer a napkin, anticipating Christiana's potential outburst.

Christiana had informed Gia about Bash's illness and subsequent death, but Natalie hadn't bothered to relay the news. Gia, who was dealing with her own diagnosis of dementia, attended the wake just as it concluded. When she walked in, Christiana made her way to her aunt to greet her. When Gia approached Bash's casket, Nadia and Lynsey assisted Natalie in standing up.

Natalie, approaching Gia, asked, "Do you remember me?" Gia, utilizing her dementia as a shield, she claimed not to know her. Gia saw no reason to engage with Natalie.

Later that evening, Christiana and Gia had a candid conversation at the hotel. Gia admitted she knew who Natalie was but chose not to deal with her due to her lack of decency in not informing her about her brother's illness or passing.

The following day, during the funeral, Christiana decided to sit away from her relatives to observe Bash's homegoing service without the anticipated nonsense. The minister, whose focus seemed more on attacking Christiana than honoring Bash, required correction from her. When Christiana requested one of the funeral employees for a few extra programs, she was told she could not get one because the rest were for the family members. Christiana did not say anything. It was one of her friends who clarified her relationship with Bash, prompting apologies from the funeral staff.

Remaining composed, Christiana, guided by the Most High, refrained from disruptive behavior at Bash's funeral. Christiana's sons thought for sure that their mother was going to lose it once they heard the minister indirectly saying things that they knew were directed towards Christiana.

Christiana looked like the sun had risen upon her throughout the service, handling each comment thrown her way with finesse, redirecting every nonsense back to the minister with a simple "amen." However, it seemed she needed to practice what she was trying to deliver, which was meant to attack.

Andre went to the podium, where he was giving thanks to guests for coming to his father's funeral; he then added that he also would like to thank Christiana for coming to the services. Christiana was thinking, why's he thanking me as if she had a choice to be there or not because he did not thank any of the other family members?

Once they all made it to the grave site and took their respective seats, the funeral arranger handed all family members flowers from Bash's casket and overlooked Christiana. They had even given flowers to nonfamily members who were not seated with the family until someone else spoke up, stating that they had overlooked Christiana.

One of the people decided to hand Christiana their flower, but Christiana refused it and said no, that they could keep it; she understood that Bash was not in that flower and it was nothing that would upset her. Once the funeral was over and they went to the repast, Christiana decided to make plates to help feed the elderly instead of sitting with her immediate family.

The minister's father walked past Christiana as she looked directly at him. Still, apparently, it made him feel some way because he foolishly said, "It was nice to meet you." To which Christiana sarcastically replied, "Oh, we met?" However, he kept moving.

Once the repast was over and everyone was about to leave, Christiana had gotten into the vehicle with her sons. She said, "I don't know about you all, but I have had about all I can handle with the rest of them, so please take me back to your place, Cassius."

Christiana refrained from visiting Natalie for the remainder of her stay, and their communication only resumed after Christiana returned to Los Angeles. Throughout this period, Natalie persistently called Christiana to share updates about Nadia's ex-husband Damon's ailing father was sick and that he was not expected to make it. Christiana would immediately change the subject because she could not believe that her mother had the nerve to keep calling her to inform her of someone else's father when she never called her regarding Bash.

Christiana understood that because it had to do with Nadia, her mother felt Christiana needed to know. The day was coming when Natalie would call Christiana and would get exactly what she was looking for.

Damon's father had passed away, and on the day of the funeral, Natalie decided to call Christiana with the particulars of it. It was then that the Most High told Christiana to convey a message to her mother. Natalie sounded somber for Nadia, who again was no longer even married to Damon, now married to his first love, but still desired a connection with Damon, at least in her mind, even after the death of his father.

Christiana told her mother, "I'm sorry for Damon's loss, but it has nothing to do with me!" Natalie sarcastically asked, "What is your problem?" That's when Christiana expressed her frustration, saying, "Let me tell you what my problem is! It's amazing how you're calling me about someone else's sickness and now the death of someone else who has nothing to do with me. You didn't bother to call me about my own father. None of you who were there also did not bother to inform me, even though you all knew he wasn't expected to survive. You all have cell phones, and there's no excuse for not letting me know."

Christiana continued, "Apparently, you forgot you have a third child, even though you called the other two."

Natalie began lying on Christiana to Christiana, and that is when the shit hit the fan.

Natalie kept lying and said, "The reason we did not call you was because the last time that we spoke with you, you went off on us."

Christiana felt as if she had gone through the roof at that point and was livid! Christiana said, "You are unbelievable!" Natalie replied, "Yeah, because you said you were sitting on your front porch, and then you began going off on us." Christiana retorted, "Well, first off, that is a lie! I was at the car dealership when you called, and it was you who first said your father does not want to talk to you any longer. All I asked was, 'What?' Besides, I heard my father crying his heart out before you made him get on the phone and forced him to tell me that he no longer wanted to talk to me. You should be ashamed of yourself."

Natalie said, "I am not ashamed of anything!"

Christiana replied, "Well, you should be, but that's between you and the "Most High" now! You didn't realize that you helped to kill him even faster by your actions in his last days, and his heart could not take it."

Christiana was about to maim her mother and remind her of all her indiscretions with different men while she was still married to Bash. Christiana was about to let her mother know that they all know that Nadia is not Bash's child. Christiana understood the real reason that her mother had an issue with her and the reason that Bash had to set his own daughter aside to prop up Nadia. It was to cover Natalie's shame, which she also understood. She was a reminder of her mother's indiscretions because you could see that Christiana looked like Bash.

Right then, the Most High sent the Jehovah's Witness to Natalie's front door. When she laid the phone down to go answer the door, the "Most High" told Christiana to hang up the phone because it was not her job to maim her mother. The Most High was going to deal with Natalie.

Christiana was only supposed to tell her mother what she needed to tell her and not all that she wanted to tell her.

Natalie tried calling Christiana again, but she did not answer her call. Christiana knew that her mother had heard everything that she was supposed to have heard and that her mother wanted to bait her into an argument. Natalie just wanted to get some things off her chest, but it would have ended badly for her. This was one conversation that Natalie did not jump on the phone to broadcast to anyone or everyone who would care to listen.

Chapter 73: On a Dime

Cassius had taken ill. He called his mother to let her know that he was about to be hospitalized, and like many other times, Christiana dropped whatever she was doing and got on an airplane to get to her son immediately. Christiana stayed with her son at his bedside for two weeks until he was released, then she returned home.

While she was there, Andre made his way to the hospital because he was feeling disenchanted with his mother. When he met with Christiana, he wanted her to take an allegiance with him by telling her how he and his mother had been tight for most of his life. Still, he came to find out that it was his mother and Nadia who were closer. Christiana listened, and once he finished talking about the relationship pecking order, that is when Christiana said, "I hear you, but amazingly, I have never even been in the 'equation,' so I don't know what to tell you." Andre looked like he did not know how to respond to what Christiana had said.

After Christiana had arrived back in Los Angeles, she was having migraines and vision issues. She soon found out that she had developed a brain tumor. Christiana ended up making a consultant appointment at the New York Presbyterian Neurology Department for a second opinion, and it was determined that she needed immediate brain surgery.

Ishmael was the only one of her sons who made the trip with his mother for the surgery. He said he did not care what was going on; he was going. Natalie asked Christiana if she needed her to go, and Christiana told her mother that if she felt she wanted to go, she could. Still, all she needed was the "Most High" to be with her. Natalie did not make any attempt to go with her daughter, nor did Cassius or Niles.

Christiana had made it through a lengthy surgery, and after the fourth day, she received a call from Cassius stating that his door had been kicked in, and they were looking for Niles for a supposed bank robbery. That is when Christiana told Cassius that she was on her way. Christiana knew that Di'Ban or none of her family members would check to make sure that her sons were alright. Christiana spoke with the surgeon and told him that she had a family emergency. She needed to go to check on one of her children.

The surgeon told Christiana that he would not advise her to leave yet, as they had previously discussed prior to the surgery, and that she needed to remain in New York for thirty days until she was released to travel. The surgeon reminded her that they were supposed to drive back to safeguard her against possible blood clots concerning flying after any surgery.

Christiana and Ishmael took a flight to Chicago the following day, and Cassius picked them up from the airport. Niles was nowhere to be found, but Christiana was seeing the news flashing her son's picture on the television. Niles finally called his mother and stated that he was not guilty of what they were claiming he had done. He said that he was giving them time to gather more evidence or information before he would turn himself in.

Christiana went and hired an attorney. She was speaking on Niles' behalf for the unsubstantiated facts that were not warranted on her son's behalf until he turned himself in to the authorities. Christiana paid hefty retainer fees for the attorney, and she was able to help decipher supposed text messages that the attorney knew would not stand up in a trial. Christiana sat in the attorney's office for many days, helping him understand her son's work schedule and other activities that he was involved in. She was also able to show where he had let an associate use his vehicle, not knowing what they were about to do. Because her son had left his gun in the car and they found marijuana, they charged him with that. However, he was sentenced to minimal time in the county jail.

Ishmael made sure that he put money on his brother's commissary, as well as his mother did. Ishmael also made sure to give Cassius extra cash in case he needed medications or food before he and his mother returned to Los Angeles. Ishmael always made sure to have his brother's back regardless of the circumstances, but time will bring about a change.

Chapter 74: High Tide

Christiana continued to fall ill after her brain surgery, and it took a toll on her. Ishmael expressed his concern to his mother, suggesting that she return to Chicago, where she would have more support than he alone could provide. He acknowledged that his financial resources were limited compared to his brothers, who could potentially look after her better.

Taking her son's advice, Christiana returned to Chicago and stayed with Niles and his girlfriend, Linda, for about two months. Despite her illness, she took on the responsibility of keeping the house clean and preparing meals as a gesture of appreciation, intending to contribute until she regained her health and secured employment.

However, tensions arose when Linda displayed discomfort with Niles spending time talking to his mother. Christiana observed Linda's desire for Niles' undivided attention and her deliberate avoidance of interacting with Christiana. This puzzled Christiana. She, therefore, remained quiet most of the time, attempting to stay out of the way of the couple until she could move out. Christiana knew that trying to divide a family over something as simple as talking with your mother would create problems of her own. Christiana decided that once she landed a job, she would ask Cassius if she could stay with him for a month to give Niles and Linda the space Linda seemed to desire.

Niles understood his mother's decision to leave. He expressed frustration with Linda's perceived nagging and complaints. He even considered moving with his mother. To support her independence, Niles helped Christiana acquire a car for her transportation needs. However, tensions escalated when Linda found out about the car. She then insisted that she should be the one driving the car despite already having other vehicles. Christiana was making payments for the vehicle. Therefore, it should not have been an issue.

Ishmael had gone to stay with his father, Di'Ban, and his girlfriend at their house. Ishmael said that it was a living hell, and all they ever did was complain about anything they could muster up about him. Christiana told her son that it would not be long before she would have a place for them. She told him just to stay away from them to keep confusion down. Ishmael said that he does, but they still come to fault-find him every opportunity they get.

Christiana had planned a move-out day from Niles and Linda's place. She put the room back into the position she had found it, leaving it clean. Christiana thanked Linda for her stay with them. Before departing, she told her that she felt that they needed their space. Linda looked like she did not know how to respond, but she shook her head to say yes.

Christiana arrived at Cassius and Nadine's place for a short stay until she could get what she needed to make the move into her own place. Cassius seemed glad to have his mother there for the visit, but with Nadine, it was a different story. In the first week, Nadine had one of her friends, apparently someone she confided in, show up at their house.

Christiana was sitting on the couch watching TV, and Cassius was in the kitchen cooking when Nadine's friend showed up. Upon entering the house, she saw Christiana, whom she had never even met. Nadine's friend wanted to show Christiana what she thought of her. She decided not to speak to Christiana upon her entry into the home.

Nadine knew that Cassius would have taken notice of it because he was able to see and hear from the kitchen. When her friend got ready to walk past Christiana, Nadine quickly asked her friend if she had spoken to Cassius' mother. She looked at Nadine and, in a nasty tone, said, "I said hi!"

Christiana thought to herself, 'This raggedy lying ass bitch knows she did not speak, but she already knew exactly where it was coming from.' The second day, the same friend came back, determined to let her arrogance lead her, but it ended up backfiring on her.

Nadine's friend had her kids with her, and they were running rampant throughout the house. Cassius told the kids to get somewhere and sit down while he was cooking, and Nadine was washing dishes. Christiana was sitting, watching a movie, minding her business, and Nadine's friend did not like the fact that Cassius had corrected her children. This friend of Nadine's came and said, "I will do whatever the fuck I want to do up in here." Christiana looked up at her, thinking, 'Oh yeah, her behind is really asking for it.' Nadine looked at her friend, trying to shake her head to tell her friend no, but without Cassius seeing it. Nadine had not opened her mouth to correct her friend, but that was when Cassius spoke up and said, "Not up in this motherfucker will you do whatever the fuck you think you want to do!" The friend then said, "Oh, you're going to jump all bad because what? Your mama is here?" Cassius told her, "As a matter of fact, get your kids with your shit and get the fuck out right now!"

Nadine cut the water off and went upstairs while Cassius continued to cook, trying to calm down. Christiana continued watching the movie because she could tell her son was upset. Two minutes later, Nadine came down the stairs with her throwaway stash overnight bag. Nadine headed for the garage, and Cassius followed her, trying to see where she was going. The next thing Nadine did was to draw Christiana into the nonsense. Nadine came back in, telling Christiana to get her son because she did not want to be there, and he was not letting her leave!

Christiana looked at her son and then asked him if he had heard what Nadine had said. Cassius said yes, and then Christiana said, "Now I want to ask you a question, Nadine. Let me get this straight. Your friend comes over to this house, she disrespects you both, and you're leaving. Is that right?" Nadine nodded, but again, Christiana turned to Cassius and said, "Well, son, she just showed you where her allegiance lies, so let her go." Christiana told her son that if someone wanted to leave, let them walk because it was clear that Nadine wanted to walk just because she already had a bag packed. His mother also told her son that she would return because she was used to what he was providing for her and her children. The following day. Christiana arrived back from work, but Nadine had made her way back, just like Cassius was warned she would do.

Christiana asked her son how much 'he needed' from her financially since she was staying there in his home, and he told her to pay nothing. He also told her she could keep her money to make her move. Christiana did not have any winter clothes since she had left Los Angeles, and the weather was about to change. Niles had taken his mother one weekend to shop for some needed winter clothing.

Christiana returned to the house with a few shopping bags while Nadine was in the kitchen. When she walked in with the bags, Nadine looked down at the bags and apparently had an issue with Christiana shopping and not paying anything to her son. The following day, Cassius suddenly asked his mother for some money, and she knew where it was stemming from, but she gave him exactly what he asked her for.

Christiana was not home very much at all because she was working a lot of overtime. She left before Cassius and Nadine woke up in the morning, and she did not return until they were in bed for the night.

Christiana diligently worked every day except for Sundays. On those days, she managed her personal errands, including grocery shopping for her work meals and taking care of her laundry. Christiana used her own laundry detergents and softeners, valuing self-sufficiency. Despite her busy schedule, she only stayed at their place long enough to rest and avoid getting in their way.

Meanwhile, Linda and Niles were facing their own set of challenges. Christiana was surprised when she received a call from Linda one day expressing a desire to meet for dinner. Niles, however, suspected Linda's ulterior motives were to seek his mother to help her manipulate him into marriage. Christiana focused on her demanding twelve-hour workdays, declined, asserting her unwillingness to be used for someone else's manipulating needs.

One Sunday morning, Cassius woke Christiana to join him for a breakfast he prepared. As she descended the stairs, she found three of Nadine's teenage daughters lounging on the couches, wrapped in blankets, watching television. After preparing the meal, Cassius took it upon himself and finished preparing plates for everyone.

Once Cassius was done cooking the breakfast and everyone had eaten, he began washing up the morning dishes; he then went to clean the remaining part of the house as they sat watching. Christiana was disturbed by what she saw once Cassius went upstairs and began cleaning the restroom that the young ladies used. Disturbed by this, Christiana spoke up, questioning why Cassius was doing everything around the house while they sat idly watching him. "Cassius, those are three big periods having females that should be cleaning their own space, not you, son."

When Nadine arrived, she decided that she was going to get with Christiana! Nadine said, "My daughters told me what you said." And that is when Christiana said, "It was no secret. I will tell you just like I told my son."

Christiana continued, "I will say whatever I want to say in regards to my son, which is you all are taking advantage of him, and there is no reason that everyone should be sitting on their behinds being waited on hand and foot around here! Now, Nadine, if you have a problem with it, then that's your problem."

Despite years of Cassius avoiding visiting his mother and his younger brother, suddenly, he and Nadine eventually went to Los Angeles, revealing the real reason behind his reluctance. Christiana, unfazed, considered it was his loss. Once they arrived back from the trip, that sounded really messed up, and once they were back into the swing of things, Nadine had family members who came to do laundry at their place each weekend. This was one of the reasons that Christiana kept her laundry supplies in the room she was staying in. Christiana had a trunk load of cleaning supplies and laundry supplies in the trunk of her vehicle because she was stocking up for her move.

Tensions rose when Nadine was upstairs in the hallway, raising hell about the use of laundry supplies. Christiana took offense to it because if she was not directing her comments to her children, then she was apparently talking to Christiana. Cassius already knew what was going on because he came down the stairs and saw his mother pick up her keys. He asked her where she was going, to which Christiana replied, "Don't even worry about it."

Cassius noticed that his mother was going to her car to retrieve some supplies to bring back into the house. He tried to stop her by saying that it was not necessary, but Christiana gave him a disapproving look. Despite this, Christiana brought two containers of laundry detergent into the house to silence Nadine.

A week later, Christiana informed her son of her decision to move and thanked him for the month-long stay. Leaving the room as she found it, Christiana was relieved to escape the chaos.

Ishmael was glad to be away from his father's house. He joined his mother for a peaceful life in their luxury apartment. However, unknown challenges awaited them, threatening the tranquillity they had come to enjoy.

Chapter 75: The Fallout

Christiana found herself working so many hours that she barely had time to keep track of the money she was earning. She had secured a luxury apartment for herself, Niles, and Ishmael, with the intention of splitting the expenses three ways. Christiana had even opened a bank account with her sons' names on it, envisioning a plan to pool their resources and pay off everyone's debts.

Her decision to take on extensive overtime, working between 65 - 70 hours a week, was driven by the goal of achieving financial stability. However, when Christiana scheduled a vacation and checked the account online to pay for it, she discovered it was unexpectedly short. The money that was supposed to be there had been nearly depleted, prompting her to question her sons about it.

In response, Christiana received nothing but excuses. She was disheartened to realize that her sons had spent the money without consulting with her. Christiana never had an issue with sharing with her children, but she realized that they were not giving her the same courtesy that she had extended to them. Forced to dip into another fund to cover her vacation expenses, Christiana decided it was time to reassess her approach to building a future with her sons.

Niles failed to follow through on the plan they had discussed, leaving Christiana with the majority of the financial burden. While Ishmael contributed what he could with his job, it wasn't enough.

Consequently, when their lease ended, Christiana and Ishmael moved into a more affordable apartment, allowing Christiana to rebuild her savings.

Despite the setback, Christiana understood that the money had been spent without her knowledge and served her no purpose in crying over spilled milk. She resolved to approach building a future with her sons differently. Little did she know that unexpected developments were about to take both her and Ishmael by surprise.

Cassius eventually decided that he was ready to purchase a home. He had discussed with his mother how he had grown tired of Nadine and her spending habits, which were keeping him in a financial hole. Cassius also stated that he was tired of her constantly accusing him of seeing other women and her stalking tendencies. Christiana told him that he would know when he'd had enough and that he would do what he needed to do for himself once he got tired.

Cassius had informed Nadine that they needed to part ways and focus on saving money because he was ready to move into his own home. Nadine pretended that she did not have any place to go, hoping that Cassius would change his mind and they could continue living in the condo they were already in. However, Cassius made plans and stuck to them.

Cassius went to Di'Ban's house to stay and save money to purchase a home. He gave his father money in the first week to stay with him, but then his father and his girlfriend decided that they no longer wanted Cassius there. Cassius arrived at his mother's home, knowing that she would not turn him away. Once he came and told Christiana that his father did not want him there, the first thing Christiana said was, "I don't understand why you went over there because he has already shown you who he is. I don't understand why you didn't come here from the start."

Cassius told his mother that his father's girlfriend said he acts too much like his mother. Christiana responded, "It's only because she has listened to what he has told her. She does not know me or anything about me to form that opinion!"

Ishmael told Cassius not to worry about what Di'Ban did not want to help with and that he could stay with them to get whatever he needed and build his money to be able to purchase a home. Ishmael also said that Cassius did not need to pay them anything, but if he or his mother ever needed anything, he knew that he would do the same for them.

Christiana suggested that Cassius could take her room and bed, and she would sleep on the couch because, being the smallest of them all, it would be more comfortable for him. Ishmael asked Cassius if he wanted him to help retrieve his belongings from Di'Ban's house.

Ishmael told Christiana that he wanted to switch out the bedrooms so that Cassius could have his own private bathroom concerning transplant concerns. Christiana and Ishmael both made no fuss about Cassius suddenly needing a place or the sacrifice they were causing other than wanting to see him achieve his dream.

A year later, after saving up, Cassius finally closed on his new home. However, that's when things were about to take a turn for the worse. Cassius had not told Nadine that he had secured a home for some reason.

He had been in his home for a few months when Christiana asked him if he had told Nadine about the house, to which he replied that he hadn't because he didn't want to drag the same drama into his new home.

Christiana told him that it should not be a secret and that it did not matter if she knew or not because they were not married, and he did not have to allow her to move in with him. Christiana told her son after about the fourth month that she knows you have your house, and she is just waiting for you to tell her so she can move in, but again, his mother said it should not be a secret.

One weekend morning, Nadine had been driving past family members' homes to figure out where he was so that she could confront him about his new home. Christiana was at home when her doorbell rang. Christiana was not expecting anyone to be visiting her, and that's when Smokey began barking and went to the front door.

Christiana went to the door, cracked it open, and saw that it was Nadine. Right then, Christiana knew what it was about because Nadine had never come to her door or visited her before. Nadine asked if Cassius was there, and Christiana politely said no, with no further questions or concerns. Nadine made a rude gesture as if she had caught him in a lie before she walked off, thinking that Christiana would have wanted to know if something was wrong!

Christiana walked away once Nadine left the door and called Cassius to tell him that Nadine was just at her door looking for him. That's when he said, "She is just trying to figure out where I am." Nadine had called Cassius and wanted him to look out the window to see if he was there. Nadine came back to Christiana's door, ringing her doorbell once more. That's when Christiana told her son to hold on because it was probably her again.

Christiana told her son she did not get the response she wanted; that's why she was coming back for more. Cassius remained on the call as his mother had told him to because she wanted him to hear the conversation for himself.

Nadine was angry, with her chest heaving like she was having a control problem. However, Christiana stayed calm while letting her son listen. Christiana then stepped out of the doorway to see what she wanted to discuss with her.

Once Nadine started talking and stating that she had had enough of Cassius, Christiana said, "I am trying to understand why you're here at my home with this because I'm not in the relationship."

Nadine was trying to figure out how to respond to her question, so the next thing she thought to do was to drop a couple of tears, apparently believing that it would get a sympathetic response from Christiana. Christiana remembered falling for fakes before, and she was not interested in any of what she was saying. Nadine got even more mad; however, Christiana wasn't worried because Smokey was at the door trying to get out once he heard somewhat of a confrontation going on.

Nadine began to wish ill upon Cassius, not realizing that Cassius had been listening. Christiana would have told her about herself after she spoke ill against her son. But Christiana remained quiet for Cassius to finish listening so he could understand her thought process and make his own decision without her putting words in Christiana's mouth, as she had done before. Cassius eventually made the decision to walk away from the relationship, which surprised his mother.

Cassius had always called his mother daily, even sometimes several times a day, for motivation or to gain wisdom and knowledge from her. Cassius told his mother that he had gone to a funeral and that it sounded like they were speaking about his mother when he talked about her character and all that she had given, leaving a legacy for her son. Cassius also said that he had mentioned something that he had often heard his mother say, which was, "Motherhood does not come with a manual.

Chapter 76: Unexpected Courtesies

Ishmael informed his mother that they needed to vacate their residence within a month. He suggested that she should speak with Cassius again since it shouldn't be an issue for his mother to stay with him, especially after they had accommodated Cassius previously. Christiana anticipated some good news as Cassius paid a visit and brought a gift for his mother. While she couldn't quite decipher the gift's meaning, she was always grateful for their gestures. Shortly after, Niles arrived and presented his mother with a card, which left her perplexed as it referred to her as "your grandson."

Upon rereading the card, Christiana had a sudden realization – she might be on the verge of becoming a grandmother! Her eyes widened with excitement as Cassius and Niles stood nearby. "Am I going to be a grandma?" she exclaimed. Niles confirmed the news, and Christiana couldn't contain her joy, even jokingly telling her dog, "Well, now it looks like you got to go" (laughing)!

Niles informed his mother that the baby would arrive in three months. Christiana couldn't understand why she hadn't heard about the pregnancy sooner. She hadn't even seen Linda pregnant, and as a matter of fact, she hadn't seen Linda in five years since she had moved out of their home.

A week later, Niles informed his mother that the baby wouldn't be visiting her house due to the presence of the dog. Christiana reminded

Niles of her earlier comment about the dog but assured him that she wouldn't put her grandchild in any danger.

Due to the pandemic, Christiana assumed she wouldn't get to see the baby for a few months. However, once Linda gave birth, Niles called his mother to visit them in the hospital. Christiana eagerly arrived at the hospital, excited to see her son holding his child and witnessing Niles' expressions.

Upon arrival, Christiana arrived at the hospital; she was having difficulties being let to go up to visit. She called Niles, who came down to the station where his mother was waiting. Christiana thought, let me ask if Linda's parents had visited or were coming because it was their daughter who had just delivered. Christiana did not want to overstep her parents when it came to their daughter.

Niles reassured his mother, informing her that Linda's parents would visit the house the following day after Linda and the baby were discharged. Christiana followed his advice, her heart pounding with excitement at the thought of meeting her first grandchild.

Upon entering Linda's hospital room, Christiana greeted her with a "Congratulations," but she received a cold, snarling fling of the hand and a cool Hey! Christiana looked and saw her granddaughter in her hospital crib. However, Christiana noticed Linda's lack of excitement and her desire not to interact with her.

Christiana badly wanted to spin on her heels and say nope, I'm good as she felt the negative vibe immediately, but Christiana did not want

Niles's moment of excitement as a new father to be stolen from him! Christiana sucked it up and stayed for a short visit while Niles had left the room, then went downstairs to talk with Ishmael.

Throughout her visit, Linda only made little effort to engage in conversation if prompted by Christiana. After enduring thirty minutes of chilly treatment, Christiana decided that she had enough, feeling undeserving of such treatment.

Christiana returned downstairs with Niles and Ishmael. His mother did not mention why she had left the visit so soon. As they left, Ishmael noticed that his mother was quiet. He asked her what was wrong, and she told him about the hospital visit.

Christiana wanted to visit her new grandchild every day, but she did not want to interrupt their bonding time. She would text Niles to check on them and ask him to let her know when she could come to see his daughter.

It took four weeks before Christiana had the opportunity to see her granddaughter again. She waited until close to noon before calling to check if it was still a go as they had previously done because she also understood that they could be tired with a newborn, so she wanted to check before going to visit them.

Niles told her she could come within an hour. When she went to get ready, he called her back, but she missed the call. When she called him back, Niles seemed to have forgotten why he had called.

Christiana asked if it was still okay to come for a visit because if she needed to come at another time, she would wait, but Niles then said, "No, it's fine to come for the visit."

Once Ishmael and Christiana arrived for the visit, it took a few minutes before anyone came to the door. When Niles finally opened the door, it was clear that he was tired and had been deeply asleep. Christiana asked Niles if he was asleep, and she figured that the entire house might be asleep. Christiana then suggested that she could come another day, but Niles again insisted that they come in.

As they walked in, they saw that Linda was sitting in a room fully awake and close to the door where she could have answered it. Christiana excitedly spoke, and again, she was shot with the driest hi that sounded like Linda could have choked on it!

Christiana understood that it was intentional once it was done for a second time. Again, Christiana wanted to leave but stayed because Niles insisted that they stay for a while. She knew he did not want to disappoint his family.

Later that evening, Christiana texted Niles, asking what the problem was and why she was not made to feel welcome. She did not want any negative or uncomfortable vibes around her granddaughter, even if it meant staying away. She told her son that she had noticed the negative attitudes both times at the hospital and at their home. The next time Christiana was allowed to see her grandchild, she was five months old. She did everything she could to minimize stress during her visit, such as wearing a mask and bringing along a clean top in case it was necessary to change.

Christiana had grown tired of feeling like she always had to schedule appointments just to see her grandchild. She realized it was up to her son to ensure that she had regular access to her grandchild. She also understood that her son had entrusted her with the care of his close friends' children in their time of need, demonstrating his mother would take care of them; however, she couldn't shake the feeling as if she was being inadequate concerning her own grandchild, which saddened her.

Chapter 77: The Disconnect

Christiana lamented that she had missed out on many of the typical experiences of grandmotherhood: holding her grandchild on her hip, rocking her to sleep, feeding her, observing her personality develop, learning her preferences, changing her diaper, and witnessing her first steps or crawl however the one courtesy that she was glad for were the few moments she did get to spend with her.

Christiana had never imagined herself becoming a grandmother, at least not in her lifetime. So, she took the initiative to write a love letter to her future grandchildren, which she carefully placed into her Bible. She hoped that if she didn't survive to see their birth, they would know how much she longed to shower them with love, just as she had done with her own children.

Grateful that she hadn't grown too attached to her only grandchild, Christiana understood the circumstances that kept her apart from her. She accepted the situation, knowing that her granddaughter didn't know her, nor did she know her granddaughter, which somewhat eased her pain. She decided not to impose herself on anyone's life and resigned herself to the idea of watching her granddaughter grow through pictures or videos. Christiana stopped subjecting herself to begging for visits, no matter the excuses.

Chapter 78: Life Lessons and Blessings

Cassius had started seeing Ava, a relative of one of his old friends. Christiana knew of Ava from before and was aware of Cassius' fondness for her even before they started dating. What Cassius didn't know was that mentioning Ava to his mother always brought a light to his face. One evening, Cassius called Christiana and panned his camera around the room so she could see Ava.

Christiana and Ava exchanged pleasantries and had a light conversation before ending the call. The next morning, Cassius called his mother, bubbling with excitement like never before. Christiana, too, felt joy for her son.

Before getting involved with Ava, Christiana advised Cassius to ensure he was truly done with Nadine due to his history with Ava and her family. Cassius assured his mother that he respected their relationship history and wouldn't let anything go wrong.

After that, Cassius called his mother several times a day, sharing his excitement about Ava. Christiana reminded him that the "Most High" had blessed him not only with a new home but also with a healthy relationship and urged him to count it all as joy.

Christiana asked Cassius if Ava was spending a lot of time with him, and upon hearing that she was, she requested her son to give Ava her phone number in case of an emergency. Ava called Christiana the next day for a refreshing conversation, unlike any she had with Cassius' previous partners.

Cassius called his mother to ask if she minded if Ava went with them for the scheduled biopsy that he needed to have. Christiana told him that she did not have a problem with it, but she wanted to know why he would ask her that. Christiana stated that since he was a grown man if he wanted her there, it was his decision. But that was when Cassius told his mother that Ava wanted to make sure that his mother was okay with her going out of respect.

A couple of days passed, and it was time to go to the hospital for Cassius to get a biopsy done on his liver. Once Cassius and Ava came to pick up Christiana, they rode and enjoyed laughter together, and that was when Christiana was about to see Ava's character.

Once they arrived at the hospital and got into the patient's room, the nursing staff came in to start prepping Cassius for the procedure. Ava, without question or hesitation, dragged her seat to the foot of Cassius's bed. She softly began rubbing his feet and legs as if she was trying to soothe him. Christiana paid attention to how attentive Ava was with Cassius.

After Cassius was taken for the biopsy, both Ava and Christiana left to grab a bite to eat. They seemed very much in tune with each other! Ava told Christiana that she wanted to take Cassius something to eat for when he was done with the procedure. Again, Christiana just paid attention to how attentive Ava was towards her son. She was blown away by Ava's concern for Cassius. His mother then realized that Ava was in love with Cassius and hoped her son felt the same way towards Ava.

Ava and Christiana decided to walk back to the hospital from the cafeteria. While they were walking and talking, Christiana found out that the relationship was more serious than she knew! Christiana was told that Cassius and Ava were talking about marriage and having a baby as soon as possible. Christiana was also told that Cassius had mentioned to Ava that his mother would be quite happy because she was not able to have the relationship that she had hoped for with her first grandchild. He then said how long his mother had been waiting to become a grandmother.

Once the ladies made it back to the hospital, Cassius was brought back from the procedure. That was when Ava sat at his bedside to literally "spoon-feed" Cassius, but he seemed to have a problem with her feeding him. Christiana did not understand why he had a problem with it or if it was because he did not like that his mother had seen him being spoon-fed. However, his mother would later address it with him.

On the ride back to drop off Christiana, it seemed like a very light-hearted ride. Cassius and Ava were talking about getting new cars, and that was when Ava said that they should get a minivan to be able to travel with the kids and that they could take Christiana with them to be with her grandchildren.

The next day, Ava called to ask Christiana if she would like to go to breakfast. The ladies sat and talked while enjoying breakfast as Ava doted on Cassius! Ava said how excited she was to have him because she had a few failed relationships and that Cassius and Ava had a history already, so she believed that would make them solid. She added that she trusted him because of their history with her family.

Ava told Christiana that she had met Ishmael when he had to come to her and Cassius' rescue one night. Ava said the vehicle they were riding in broke down, and they were not able to reach anyone else. Nevertheless, as soon as they called Ishmael, he came within minutes from them, placing the call to him. He helped to get his brother's vehicle home safely.

Christiana also shared some of her past life experiences with Ava. To her understanding, Christiana understood how Ava's life had gone through her own life experiences, which Christiana was about to learn more on a deeper level from Ava.

Chapter 79: Revealing Traits

Christiana had suffered four strokes, which had made her unable to work. She was on long-term disability from her employer due to illness, but it had run out. Meanwhile, Cassius had been living with them for a year, but he had not been asked to contribute anything to help the family. Instead, he had been able to save up enough money to purchase himself a nice home.

Ishmael was still working, but the situation had become more difficult since Christiana lost her income. The strain of trying to hold everything together was taking its toll. Ishmael had recently bought himself a nice car with Christiana's help before her income status changed. The car payment and insurance added to his financial pressures, along with all the other bills.

Ishmael asked his mother if she could stay with Cassius for a short time, and he would stay with a friend while Ishmael could save up enough money to move them into a cheaper apartment.

One day, Christiana asked Cassius if she could stay with him for a while. Ishmael could save enough money for them to move into a more affordable place. Cassius looked like his mother had asked him for his last breath; he jumped up and said that he would be right back, as he needed to run to the store. Cassius left the house without answering Christiana's question. He returned with a jug of Hawaiian punch, but he did not answer his mother's question or offer any help in their time of need.

Ava stayed in touch with Christiana and wanted to build a relationship with Cassius. She thought it was essential to have a good relationship with a mother figure in her life and wanted to create a healthy family. Christiana appreciated Ava. She believed that Ava was about to aid in bringing change to her family because she had a way of drawing people!

Christiana recognized that Ava valued a unified family structure, but her two sons apparently valued people based on their status. Despite knowing the truth about people who had come against them, they overlooked it based on earthly possessions. Cassius suggested to his mother that if any madness were to take place in the world, the meet-up place would be at Di'Ban's house. However, Christiana refused, as she believed that her son had to be out of his mind to suggest that she should go in the last days to take a seat in a camp with a serpent.

Christiana told her son, no, thank you, but that she would be where she would be if anything were to jump off! Christiana told Cassius she would not be there because people don't change. They just change up how they do things. Christiana knew exactly what she was talking about, and it was going to be revealed very soon! Christiana gave her sons more credit than she believed them to have, and she believed them to be wiser men as well. However, she was about to be reminded of a comment that she was once told would happen to where that blood would come out. Cassius must have wished that his parents would get back together, as he told his mother that his father was thinking of calling her to apologize. However, Christiana understood she did not need any apology from Di'Ban, as she also knew she did not possess any heaven or hell to place him in. The first thing that she had already done was to apologize to herself for dealing with Di'Ban! Cassius told his mother that she needed to let go of her resentment, but Christiana replied that she did not possess any grace or mercy to give to Di'Ban and that he should not think that he had an option to swoop back around to pick her up.

It was only a couple of weeks before Ishmael and his mother had to move out of their dwelling when Christiana contacted her son Cassius once again. She let him know that she had no place to move to and asked if she could come there for a while. Cassius became irate once more at the idea of his mother needing to stay with him for any length of time. He told her no, stating that he needed his place to himself.

Christiana had made other calls, one of which was to Manius. He told his aunt that she could come there. However, Manius and his girlfriend were already experiencing relationship problems, and Christiana didn't want to add anything else to their plates. Ishmael had called Tony to see if he could stay with him, and Tony agreed. However, Christiana knew this was a bad idea due to Tony's bipolar personality.

Ishmael was already feeling depressed because his two older brothers weren't trying to help him or his mother. Christiana watched as Ishmael sat and cried, hurt by the only two people other than his mother, whom he considered family. Ishmael said he couldn't believe his brothers because they had never had to ask him or their mother for help when they needed it.

He realized that he and his mother had always gone to great lengths for his brothers, but his brothers wouldn't do the same for them. Ishmael expressed frustration at being used as party fillers for the ones who did nothing but wipe their asses on them but then acted as if they have memory loss towards those same people and how they were quick to want to drag him with them for the bullshit. Ishmael told his mother that he does not remember them even bringing him a cupcake for his birthday or celebrating their mother, but was quick to celebrate other women and their families.

Christiana was about to sacrifice herself again out of concern for Ishmael's mental state. She planned to go with him to Tony's to help ease the stress on her son while they stayed there.

Suddenly, the day before they were scheduled to move in, Tony told Ishmael that Smokey couldn't come because they didn't allow aggressive dogs in the neighborhood. Christiana was relieved that Tony revealed his true colors before they moved in even a single bag because nothing about him had changed.

Ava had told Christiana that she could come to stay with her rather than go to a shelter. However, Ishmael still didn't have a place to stay. He had considered living in his car, but the weather in Chicago had already turned frigid. Christiana told her son he couldn't live in his car, adding to an already stressful situation. She called Cassius and told him what his youngest brother was planning to do, hoping he would lend Ishmael the same courtesy he had given Cassius for a year.

Christiana had learned some things about her character from what Cassius had told Ava. Ava told Cassius that she was taking his mother in because Christiana had no place to go, but Cassius told Ava that he did not give a fuck if they did not have a place or if they were left on a curb. Cassius told Ava he had gotten his house only for himself and that he did not want to hear shit else about Ishmael or Christiana's problems because he was tired of taking care of them. Cassius told Ava that his mother had fucked up his entire life and that she had been planning on taking over his house, but he was going to show her something new!

Cassius immediately began treating Ava differently, speaking to her disrespectfully, as if she was some shit on the bottom of his shoe. Christiana noticed that Cassius was letting this shitty ass serpent spiel poison to him and was willing to say anything to turn her sons against their mother, all for the ultimate goal of excluding Ishmael and Ava for not falling in line with the demise, but it was mainly out of pure envy. Philippe's mother, Donna, learned that Ishmael was planning to live in his car and called Christiana, offering her home to Ishmael and Smokey for their safety.

Cassius and Di'Ban both decide to drag Niles into their twisted web of lies, believing that there's more power in numbers as long as everyone agrees to say the same thing about someone while attempting to manipulate Ava for the demise. Di'Ban held onto a voicemail recording from over thirty years ago between Mykale and Christiana, twisting its meaning to further his agenda against Christiana. However, Christiana's sons believed his lies without questioning or thinking of what his purpose was. On the other hand, she believed that her sons were smarter men and that they would see through the wicked scheme. Di'Ban did not reveal that he was eavesdropping on Christiana's conversations, nor that he and Christiana were not married or living together at that time. His ego was bruised, which festered in his mind for too long. Christiana's sons believed his lies without questioning him.

Once Niles mentioned Mykale's name during a phone conversation, Christiana remembered that her sons had only met him once or twice and were too young to even remember Mykale. At that moment, she understood who it came from and what the ploy was! Christiana remembered the mad monster who was sitting and waiting for any preying opportunity to rear his ugly head just for some get-back while sporting a shit-eating grin. Little did the evil man know that he would soon fail, and he would have to eventually beg the Most High for forgiveness.

Christiana wondered if Cassius was ill and if that was why he was acting the way he was. His mother could not identify who her son was anymore! Christiana thought it was probably due to him trying to spare his mother's heart out of concern about not being able to handle the possible loss of any of her children. She knew that she would do anything for her children, even if it were any attempt to try to move heaven and hell for them, even if it meant facing challenges alone.

Chapter 80: Curbside Pickup

By this time, Christiana had taken up on Ava's offer and was living with Ava and her two children, Tyler and Breanna. Not only did Ava accept her as family, but so did her children. This actually worked out better than they both could have guessed. Christiana was able to help Ava with her children and around the house, which gave Ava a break from doing things all on her own, without any help prior to Christiana ever arriving there!

Shortly after Christiana moved in with Ava and her family, Ava told her mother that she had let Cassius' mother move in with her. That is when her mother told her that she had made a big mistake! Ava asked her mother what caused her to say that. Ava's mother told Ava that Christiana was crazy!

Now, not one time had Ava's mother ever met Christiana or even spoken with her to have come to that conclusion. Ava's mother told her that Cassius had told the rest of her family that his mother was crazy.

Christiana could not believe that her son had been telling people that she had a mental problem, but she understood that, apparently, he had had an issue with her for some time now. What Christiana did not understand was why her son would mislead people about her by tainting her character or her mental state of mind.

Cassius called Ava, and he told her that she did not know who she had let in her house and that she would find out by phase two. When Ava asked him to tell her what he meant, Cassius told her that he would do her a favor and take her to Di'Ban's house so he could tell her the truth about his mother!

Ava told Cassius that she was not interested in hearing it because she saw his mother for who she was or was and that he needed to treat his mother better. Ava also told Cassius that his mother loves him. Ava told him that she had watched his mother cry out of concern for him, and that is when Cassius told Ava he did not want to even hear the shit.

Ava tells Cassius that her children love his mother, and so does she. But if there was something that she needed to know about his mother, he needed to tell her! Cassius told Ava that his mother used to drug him and his brothers as kids just so she could go out to be with different men and that she probably was not even at work when she said she was.

Cassius then stooped lower concerning his mother and told Ava that his mother was the reason that he had to have a liver transplant! Cassius stated it was due to her drugging them with "Dimetapp" cough medicine. Once Ava told Christiana what Cassius had said about her, Christiana was in a surprisingly disbelieving state of mind with the horrible accusations that were spoken against her by both of her sons!

Ava was also told that Christiana had sat on the sideline, letting Tony mentally and physically abuse them, but again, this was the first time that Christiana had heard of anything about any form of abuse.

Christiana said that if that was the case, then she could not understand why Niles went to stay with Tony, even after she had divorced him and moved to Los Angeles. Christiana told Ava, "Yes, Tony seemed bipolar but abusive towards my children. Does not make sense. Especially when they all knew that they together would have beat Tony's ass to a bloody pulp!"

Ava was in a conversation with Cassius when she asked him about their relationship. Then, she told him that he needed to talk with his mother to fix the problems between both his mother and his younger brother! Ava told him she needed to have his mother be a part of their lives, mainly once they had a baby! But in return, Cassius told Ava that his mother would not be around his baby and that there was a reason that Niles would not let her see his daughter.

Christiana said, "I am just now learning that my sons have a problem towards me after all these years, and I never knew that they ever had this type of animosity towards me!" Christiana went on to say, "What I do know is that they just changed who I was as their mother!"

Christiana said, "That mother, who used to run up faster than anyone for them, is now gone! I love them because they are my children, but I also believe what they have said in the most malicious manner against me and have shown me, for whatever the reason."

Christiana spoke with Niles one day to try to find out what was really going on with Cassius, and little did he realize that he exposed where all the insanity was stemming from. Christiana asked her son if he had talked with Cassius or if he had noticed a change in his personality because she did not think that he was acting like himself.

Niles began attacking his mother by saying that throughout their childhood, she did not let them live their lives as they saw fit and that she kept moving them from different men's houses. Christiana asked him, "What men are you talking about? If it was Sicarii, we were only there for a week, and it was because I had left your father to keep him from putting his hands on me. It was not like you all did not know who Sicarii was!" That is when Niles made claims that Sicarii was also abusive to them, but again, it was the first time Christiana had ever heard of that.

Niles said, "That's what I am talking about. Everything is always about you." and that is when Christiana said, "Well, aren't you all talking about me? Now what? I am not supposed to defend myself?" but then she stopped and said, "I tell you what! I have given 100 percent of myself to you all unselfishly, without me even placing myself in the equation," but that's when Niles told his mother, "No, she had not!"

Right then, Christiana said, "What I am not going to do is defend my character to you all because I know what I have done, but I also know what I have not done! You all can continue to talk about me and believe whatever you all are being told or choose to believe because I don't have to convince anyone!"

Christiana understood that they were well old enough to remember the truth about her! Christiana also understood that they were just as foolish as the fool who was spoon-feeding them the bullshit. Finally, she wished them well with their agenda because she knew that none of them possessed any power to make or break her. She had already dealt with that devil and came out like a diamond defined under pressure!

Chapter 81: Camp Of Enemies

Christiana watched as Ava's heart was being torn apart by her son, Cassius. He had sold Ava a dream and then snatched it back without a just cause! Christiana was constantly remembering and had asked him if he even realized what a jewel Ava was.

Christiana also told Cassius, "One day, you will notice how attentive Ava has been towards you. You should not take it for granted because a love like that does not come very easily or often. You should be mindful that she is showing you how much she loves you."

Ava continuously tried extending olive branches towards Cassius because she could not believe that he would treat her in such a cool manner! Each day, Christiana had to watch as she went back and forth between her emotions, trying to figure out how to get Cassius and herself back to the happy place that they were once at!

Ava also understood and said that if he could treat his mother the way he was treating her, it would just be a matter of time before he would treat her the same.

Christiana was trying to figure out how to move out of the house quickly so that she would be out of the way of Ava and her children. Ava told Christiana, "She did not need to rush because if the only way that Cassius would be with her was if she turned on his mother, then that was something that she could not and would not do because that is a controlling person. Ava also said I understand that was what Nadine had done, and he had become accustomed to that negative behavior."

Christiana had not heard from either of her older sons at all; they would not call her to see if she needed anything or if Ishmael was alright! Cassius had blocked both Christiana and Ishmael from being able to call him; Niles would only respond by texting if his mother asked for pictures or video to be able to see her grandchild, and sadly, that was how she watched her grow!

Chapter 82: Divided, You'll Fall

Christiana understood the maneuverings of those who scapegoat people. What she never anticipated was that narcissistic behavior would come from any of her children because they themselves had watched their mother being scapegoated the majority of their lives, and they apparently chose that same behavior.

Once Christiana heard how they looked down on Ishmael and talked about him like he did not share the same blood that ran through their veins, their mother and Ishmael were both disheartened by their actions! Ishmael and his mother were getting ready to see through clear lenses.

Christiana was on her way to bed late one evening. On the other hand, Ava was excited that she was getting ready to go and visit Cassius that same evening. She thought that they were getting ready to get back on track. That is when Donna called Christiana, sounding hysterical. She stated that Christiana needed to get to an accident scene because Ishmael had been involved in a serious car accident. Donna was trying to keep herself calm, but she could no longer contain the anxiety and began screaming, telling Christiana, "Get there now! He is wrapped around a tree."

Ava had not pulled out of the driveway yet when Christiana flagged her down, stating that Ishmael had been involved in a car accident and that it sounded like it was serious. Ava did not hesitate or show any frustration with Christiana in her time of need. All she said to her was, "Let's go."

Christiana was feeling frantic. She remembered running to another car accident scene in which Cassius had been involved. But once she heard that there was a tree involved, she did not know what she was about to see.

Ava said, "Calm down, it is going to be alright," until they reached the scene and saw so many lights. As soon as the car stopped, Christiana jumped out of the car. She heard the Jaws of Life come on, and that's when she took off running, trying to reach the scene to find her son. Christiana got there and could not believe her eyes because the car was unrecognizable!

Christiana was afraid of seeing her son being pinned or even possibly dead! Christiana's tears were streaming down her face, and her heart was beating uncontrollably. She did not even notice that it was freezing cold! Christiana could not see Ishmael, and all she kept asking was, "Where is my son?" Once they got the person out of the vehicle and brought them to the street, the rescue squad was holding her back while asking if the person was her son.

Christiana could not tell until they let her closer to view. That was when she told them, "No, that is not my son!" She looked further down the street and saw two other vehicles. One looked like Ishmael's. However, she was not allowed to get any closer! Once more, she could not locate her son. She still did not know if he had already been transported or what had taken place.

Christiana had so many thoughts going on in her head. She even thought that *maybe her son may have tried to kill himself. Maybe from feeling the coldness coming from his brothers and the lack of concern for whether he lived or died?*

Donna and Ava helped with checking the accident scene for her son because he had not been transported or found near the scene. That's when Donna and Ava said, "He is not here," Donna said, "I am so sorry, but my daughter called me and told me that she believed that Ishmael was in the car accident. That's when I came rushing to the scene thinking that it was Ishmael."

Christiana kept calling Ishmael's phone but was unable to reach him. Ava had already texted Cassius before they had pulled out of the driveway of her home to let him know what was going on. After Christiana left the accident scene, she called, thinking that maybe Ishmael had tried to call Niles, and she wanted to see if he knew if his brother was alright!

Once Niles answered the phone call, his mother asked him if he had spoken with Ishmael. Niles opened his mouth, and all she heard was negativity with no concern! Christiana then realized that he had already heard that his brother had supposedly caused an accident and stated that he had left the scene. That was when Christiana understood that Di'Ban, Cassius, and Niles had not even bothered to leave the comfort of their warm homes (because it was around ten degrees that evening) to check the scene to see if he was even still alive.

Christiana was going to soon find out what they did. They jumped on the phone to call each family member, running Ishmael down, all based on hearsay, condemning him! Meanwhile, it was only his mother out of the family who arrived for her son to check on him!

Christiana had never gone to Di'Ban's house, and the only reason that she went then was to see if Ishmael had gone to his father's house and only because Ava suggested going to see if Ishmael was over there, but his mother said she doubted it, and when they saw that the lights were on, Ava went and knocked on the door first. Ava said that the only reason that she beckoned for Christiana to come up to the house was that she could hear Di'Ban yelling and hollering! Ava thought that Ishmael was there. Until Di'Ban opened his door, and she saw that he was on the phone running his son down to whoever it was on the other end putting on a shit show for them.

Christiana walked up near the porch, then looked up to the door, and as soon as Di'Ban saw Christiana, he went to grab his door. He was waiting for her to step upon his porch so he could slam the door in her face. Christiana just looked past him and at Ava. She shook her head, letting Christiana know that Ishmael was not there.

Christiana spun on her heels and went back to the car soon as she heard Di'Ban say, "I have always taken care of my boys, and I'm trying to figure out what the fuck Ishmael's problem is!" Di'Ban also told her that he does not go to jail and that he had not even tried to go for his other son when he was in trouble once before. Christiana said to herself *let me hurry up and get to this car* because she knew he was an opportunist trying to provoke her at the wrong time. She knew that she would have laid him and his misery down that evening for sure if she had fallen for his foolish ploy!

Christiana told Ava, "Once she got back in the car, he wanted to start some nonsense even though it was not the time for it. Because the only thing that he should have been concerned with was if his son was alright! Meanwhile, all he had done was kept his ass up in the comforts of his home to complain, as to play the victim."

Di'Ban called an attorney a couple of days later and ran down a hearsay story without even talking to his son to check on him or to get any facts! Di'Ban told his attorney, "Ishmael was drunk, traveling at a high rate of speed, that had caused an accident, and then left the scene." while telling the attorney, he also said, "He needs to turn himself in," while apparently thinking that he knew it all! What he did not know was that he had created an issue for his son because a lawyer has a duty as a mandated reporter in any such situation while running his mouth with no facts.

Cassius called Ava to come over to his father's house because they supposedly had something that they wanted her to tell Christiana. Turned out that it was about them calling the attorney and the story that they had told him. Ava said, "Christiana found Ishmael, and he was apparently okay. His car had been stolen, which he had reported!" Ava was telling them about how scared Christiana was walking upon the scene. Di'Ban immediately came with this bitter attitude about his son, stating that Ishmael would have been "fucked," all because his mother had shown up looking for her son, thinking that she knew everything. Also, Christiana needed to learn to stand back and be quiet!' Ava said he sounded like a proud boy with that stupid comment about how "Christiana thinks she knows it all!"

Ava told Christiana about what they were saying about Ishmael, stating, "Yeah, he's going to prison with his stupid ass." That is when Christiana said, "They did not even bother to go see if he was in an accident or if he was alive, but I also understand that is what a parent does and what a narcissist also does." Ava said, "It seemed as if Di'Ban was more interested in trying to impress me about himself, directly in front of Cassius, while up in my face, which I found to be utterly disrespectful and sickening."

Chapter 83: Bent But Not Broken

Ishmael's brother's behavior that night changed him. His heart wanted to harden towards his brothers. But his mother kept talking with Ishmael, letting him know that *all things happen for a reason*. Christiana told her son, "The Most High needed them both to see the truth about the people who are the closest to them, which are family because they can hurt you like no one else. And there's also a lesson in it!" Christiana believed the lesson was because they both used to run towards each other, but the Most High wanted to show them who wouldn't run for them! Christiana told him, "We love them, yes. But from now on, we should only run to help if we can help instead of going out of our way to help. Once after, we were shown that we were not even offered the breadcrumbs from their tables. But against all odds, we are still alive!"

Final Readers Note

Scapegoats are chosen for the reason of selfishness, consumed gain, and delusional behaviors, but most importantly, you should pray for endurance against all odds! This is for those who bruise your heels; just know that they will eventually come to their own demise, and then humility follows!

The End!

Author Cecelia Rose

Made in the USA
Monee, IL
10 May 2025

17202258R00256